STECK-VAUGHN

# Building Strategies for GED Success

## Mathematics

www.Steck-Vaughn.com
1-800-531-5015

## STAFF CREDITS

Design: Amy Braden, Deborah Diver, Joyce Spicer
Editorial: Gabrielle Field, Heera Kang, Ellen Northcutt

## ILLUSTRATION

Franklin Ayers pp.18, 20, 21, 25, 71, 74, 75, 105, 109, 145, 146, 147. All other art created by Element, LLC.

**ISBN 1-4190-0798-X**



Printed in the United States of America.

5 6 7 8 9 10 1689 15 14 13 12 11 10

4500246580

# Contents

# To the Learner

Congratulations! You have taken an important step as a lifelong learner. You have made the important decision to improve your mathematics skills. Read below to find out how Steck-Vaughn *Building Strategies for GED Success: Mathematics* will help you do just that.

- Take the **Pretest** on pages 3–7. Find out which skills you already know and which ones you need to practice. Mark them on the **Skills Preview Chart** on page 8.

- Study the five units in the book. Learn about place value, whole numbers, decimals, percents, fractions, ratios, and proportions. Check out the **GED Tips**—they have a lot of helpful information.

- Complete the **GED Skill Strategy** and **GED Test-Taking Strategy** sections. You'll learn important reading, thinking, and test-taking skills.

- As you work through the book, use the **Answers and Explanations** at the back of the book to check your own work. Study the explanations to have a better understanding of the concepts. You can also use the **Glossary** on page 150 when you want to check the meaning of a word.

- Review what you've learned by taking the **Posttest** on pages 145–148. Use the **Skills Review Chart** on page 149 to see the progress you've made!

# Setting Goals

A goal is something you want to achieve. It's important to set goals in life to help you get what you want. It's also important to set goals for your learning. So think carefully about what your goal is. Setting clear goals is an important part of your success. Choose your goal from those listed below. You may have more than one goal. If you don't see your goal, write it on the line.

My math goal is to

- get my GED
- improve my job skills
- get a new job that uses math

---

A goal can take a long time to complete. To make achieving your goal easier, you can break your goal into small steps. By focusing on one step at a time, you are able to move closer and closer to achieving your goal.

Steps to your goal can include

- understanding math vocabulary
- understanding numbers and graphs in the newspaper
- managing money better
- understanding numbers on TV news and sports
- helping your children with math homework

We hope that what you learn in this book will help you reach all of your goals.

Now take the *Mathematics Pretest* on pages 3–7. This will help you know what skills you need to improve so you can reach your goals.

# Mathematics Pretest

This *Pretest* will give you an idea of the kind of work that you will be doing in this book. It will help you to figure out how well you understand math skills and which math skills you need to improve.

You will solve different types of math problems based on the skills in this book. There is no time limit.

## Place Value

**Write the place value of each underlined digit.**

**1.** 4,570 ________ **2.** 691 ________ **3.** 211,084 ________

**Write each of these numbers.**

**4.** 5 tens 4 ones ________

**5.** 3 hundreds 1 ten 8 ones ________

**6.** 2 thousands 7 hundreds 9 ones ________

**Compare each set of numbers. Write > for *greater than* or < for *less than*.**

**7.** 7 ☐ 5 **8.** 463 ☐ 643 **9.** 82 ☐ 59

**Round each number to the nearest ten.**

**10.** 655 ________ **11.** 343 ________ **12.** 107 ________

**Answer the word problems.**

**13.** Last week, Elizabeth worked 42 hours and Barbara worked 35 hours. Who worked more hours last week?

**14.** A local factory hired 585 people. Rounded to the nearest ten, how many people did the factory hire?

## Adding and Subtracting Whole Numbers

**Solve.**

**15.** $8 + 5 =$

**16.** $17 - 8 =$

**17.** $7 + 6 =$

**Solve.**

**18.** What is eight and seven more?

**19.** What is nine less than sixteen?

**Solve.**

**20.** $\begin{array}{r} \$53 \\ +\ 16 \\ \hline \end{array}$

**21.** $\begin{array}{r} 64 \\ -\ 34 \\ \hline \end{array}$

**22.** $\begin{array}{r} \$475 \\ -\ 248 \\ \hline \end{array}$

**23.** $\begin{array}{r} 392 \\ +116 \\ \hline \end{array}$

**24.** $\$36 + \$582 + \$80 =$

**25.** $5{,}008 - 725 - 1{,}140 =$

**Round each number to its lead digit. Rewrite the problem. Then add or subtract.**

**26.** $\begin{array}{r} 86 \\ +\ 24 \\ \hline \end{array}$

**27.** $\begin{array}{r} 307 \\ -\ 188 \\ \hline \end{array}$

**28.** $\begin{array}{r} 785 \\ +\ 592 \\ \hline \end{array}$

**Answer the word problems.**

**29.** One weekend 32 men and 39 women volunteered to plant flowers in a city park. How many people volunteered all together?

**30.** There were 250 tickets sold for a school play. If 12 people did not attend the play, how many did attend?

## Multiplying and Dividing Whole Numbers

**Multiply.**

**31.** $\begin{array}{r} \$123 \\ \times \quad 3 \\ \hline \end{array}$

**32.** $260 \times 25 =$

**33.** $\begin{array}{r} 1{,}403 \\ \times \quad 46 \\ \hline \end{array}$

**Divide. Use multiplication to check your answers.**

**34.** $9\overline{)351}$

**35.** $287 \div 7 =$

**36.** $6\overline{)306}$

**Round each number to its lead digit. Rewrite the problem. Then multiply or divide.**

**37.** $37 \times 58 =$

**38.** $7{,}820 \div 23 =$

**39.** $396 \times 51 =$

**Answer the word problems.**

**40.** A charity pancake breakfast raised $3,052 by selling 763 tickets. Each ticket cost the same amount of money. How much did each ticket cost?

**41.** Leon works 8 hours each day. Last month, he worked 21 days. How many hours did Leon work last month?

## Decimals and Percents

**Solve.**

**42.** $3.4 + 2.5 =$

**43.** $9.5 \times 13 =$

**44.** $1.9 - 0.4 =$

**45.** $0.2\overline{)4.06}$

**46.** $\begin{array}{r} \$27.17 \\ -\ \ \ 0.74 \\ \hline \end{array}$

**47.** $1.5\overline{)9.015}$

**48.** $\begin{array}{r} 3.04 \\ \times\ \ 4.6 \\ \hline \end{array}$

**49.** $\begin{array}{r} \$2.85 \\ +\ 0.81 \\ \hline \end{array}$

**Round each decimal to the underlined digit.**

**50.** $5.\underline{3}6$

**51.** $22.01\underline{4}$

**52.** $9.\underline{8}83$

**53.** $12.\underline{9}4$

**Solve.**

**54.** 20% of 60 is what number?

**55.** What percent of 24 is 42?

**56.** 220 is what percent of 550?

**Answer the word problems.**

**57.** Ann bought a package of 12 plastic spoons for $1.49. Rounded to the nearest cent, what is the cost of each spoon?

**58.** Charles saves 15% of his $390 weekly salary. How much money does Charles save each week?

## Fractions, Ratio, and Proportion

**Write equivalent fractions.**

59. $\frac{3}{12} = \frac{\square}{4}$

60. $\frac{4}{5} = \frac{\square}{25}$

61. $\frac{7}{8} = \frac{21}{\square}$

62. $\frac{4}{30} = \frac{2}{\square}$

**Solve. Change improper fractions to mixed numbers. Reduce if possible.**

63. $\frac{5}{12} + \frac{1}{12}$

64. $5\frac{3}{4} - 2\frac{2}{7}$

65. $\frac{3}{5} - \frac{1}{6}$

66. $4\frac{1}{8} + 1\frac{2}{3}$

67. $\frac{1}{8} \times \frac{2}{3} =$

68. $7\frac{1}{3} \div 9 =$

69. $4 \times 2\frac{3}{4} =$

70. $15\frac{3}{4} \div 5\frac{1}{4} =$

**Solve for *n*.**

71. $\frac{4}{n} = \frac{8}{14}$

72. $\frac{2}{3} = \frac{6}{n}$

73. $\frac{1}{8} = \frac{n}{40}$

74. $\frac{n}{6} = \frac{12}{24}$

**Answer the word problems.**

75. Sharise walks $3\frac{1}{2}$ miles per day for 5 days each week. How many miles does she walk all together in 1 week?

76. A 32-ounce bottle holds 4 servings. How many servings are in a 48-ounce bottle?

**When you finish the *Mathematics Pretest*, check your answers on page 151. Then look at the chart on page 8.**

## Skills Preview Chart

This chart shows you which mathematics skills you need to study. Check your answers. In the first column, circle the number of any question you missed. Then look across the row to find out which skills you should review as well as the page numbers on which you can find instruction on those skills.

| Questions | Skill | Pages |
|---|---|---|
| 1–14 | Whole Number Place Value | 10–17 |
| 15, 17, 18, 20, 23, 24, 29 | Whole Number Addition | 28–31 |
| 16, 19, 21, 22, 25, 30 | Whole Number Subtraction | 34–37 |
| 31–33, 37, 39, 41 | Whole Number Multiplication | 54–59 |
| 34–36, 38, 40 | Whole Number Division | 60–64 |
| 26–28 | Whole Number Rounding and Estimation | 44–45 |
| 42, 49 | Decimal Addition | 84 |
| 44, 46 | Decimal Subtraction | 85 |
| 43, 48 | Decimal Multiplication | 86–87 |
| 45, 47, 57 | Decimal Division | 88–90 |
| 50–53 | Decimal Rounding and Estimation | 82–83, 91 |
| 54–56, 58 | Percents | 96–97 |
| 59–62 | Equivalent Fractions | 113–115 |
| 63, 66 | Fraction Addition | 120–123 |
| 64, 65 | Fraction Subtraction | 124–127 |
| 67, 69, 75 | Fraction Multiplication | 130–131 |
| 68, 70 | Fraction Division | 132–133 |
| 71–74, 76 | Ratio and Proportion | 135–137 |

# Unit 1 Whole Number Place Value

round

plus

value

digit

**In this unit you will learn about**

- identifying whole number place values
- finding place values for whole numbers
- forming numbers from place values
- renaming whole numbers
- comparing whole numbers
- rounding whole numbers to the nearest ten

You can see whole numbers all around you. They appear on signs, in newspapers, in stores, and in many other places.

You use whole numbers every day. You might keep score in a game or give your street address to someone. Write a whole number from your life, such as your age or the number of people in your family.

______________________________

Whole numbers can have one digit or many digits. For example, you might drive 5 miles to the mall or fly 4,300 miles on an airplane.

Write a whole number with more than three digits, such as the population of the place where you live or the year you were born.

______________________________

# Whole Number Place Names and Values

Whole numbers are made up of digits. A **digit** is one of the ten symbols—0, 1, 2, 3, 4, 5, 6, 7, 8, 9—used to write numbers. For example, the number 325 contains three digits—3, 2, and 5.

Each digit in a number has a different **place name**. The first three place names are the ones place, the tens place, and the hundreds place. **Place value** is the value of a digit based on its position in a number.

Look at the number 325 in the chart. The digit 5 is in the ones place. Its value is 5. The digit 2 is in the tens place. Its value is 2 tens, or 20. The digit 3 is in the hundreds place. Its value is 3 hundreds, or 300.

Sometimes the digit 0 (zero) appears in a number. Look at the chart again. In the number 302, the digit 0 is in the tens place. The number 302 has 0 tens.

| hundreds | tens | ones |
|---|---|---|
| 3 | 2 | 5 |
| 3 | 0 | 2 |

**Write each number in the chart.**

1. 64
2. 739
3. 87
4. 406

| | hundreds | tens | ones |
|---|---|---|---|
| 1. | | 6 | 4 |
| 2. | | | |
| 3. | | | |
| 4. | | | |

**Write the place of each underlined digit.**

5. $\underline{5}4$ = tens
6. $\underline{1}63$ = ______
7. $9\underline{8}9$ = ______
8. $\underline{8}32$ = ______
9. $1\underline{4}$ = ______
10. $\underline{4}0$ = ______
11. $6\underline{0}7$ = ______
12. $53\underline{0}$ = ______
13. $49\underline{9}$ = ______
14. $\underline{7}16$ = ______
15. $\underline{4}$ = ______
16. $1\underline{7}$ = ______

**Check your answers on page 151.**

Models of groups of ones, tens, and hundreds can be used to show the value of a place.

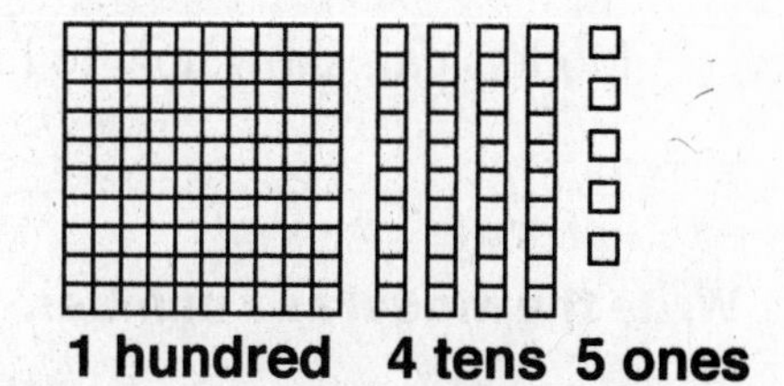


**Think about the digits and the places in the number 145.**

**Write 223 as groups of hundreds, tens, and ones.**

**1.** The digit 2 is in the hundreds place. The hundreds model below shows 2 groups of hundreds, or **2 hundreds**.

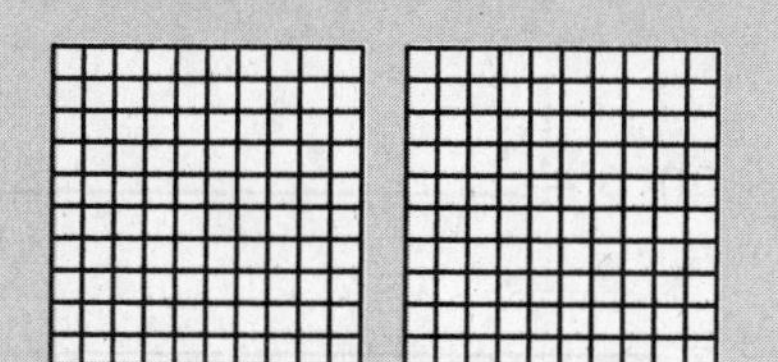

**2.** The digit 2 is in the tens place. The tens model below shows 2 groups of tens, or **2 tens**.

**3.** The digit 3 is in the ones place. The ones model below shows 3 groups of ones, or **3 ones**.

**Write the place value shown by each group of models.**

**17.** 

2 hundreds

**18.** ______

**19.** 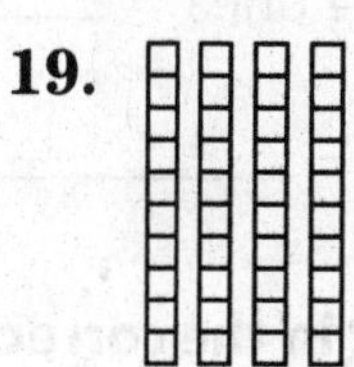

______

**Write the place value of each underlined digit.**

**20.** 624 = 2 tens

**21.** 57 = ______

**22.** 280 = ______

**23.** 9 = ______

**24.** 993 = ______

**25.** 40 = ______

**26.** 304 = ______

**27.** 188 = ______

**28.** 17 = ______

**29.** 67 = ______

**30.** 320 = ______

**31.** 588 = ______

Check your answers on page 152.

# Finding Place Values for Numbers

Models can show how to break apart the place values of a whole number.

| Write the model as a number. | Write the model as a number. | Write the model as a number. |
|---|---|---|
| 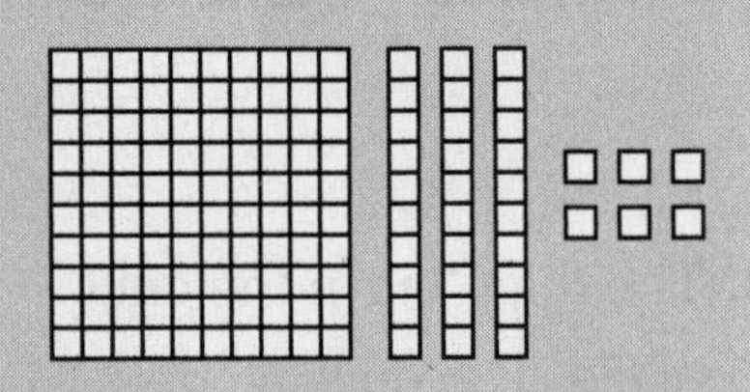 |  | |
| 1 hundred 3 tens 6 ones = **136** | 2 tens 3 ones = **23** | 6 ones = **6** |

**Write each of these place values as a number.**

**1.** 3 tens 1 one = **31**

**2.** 2 hundreds 4 ones = ________

**3.** 9 ones = ________

**4.** 4 tens 8 ones = ________

**5.** 7 hundreds 7 tens 3 ones = ________

**6.** 5 tens = ________

**7.** 9 tens 4 ones = ________

**8.** 6 hundreds 8 tens 6 ones = ________

**9.** 7 tens = ________

**Write each digit in the correct place.**

**10.** 85 = **8** tens
**5** ones

**11.** 321 = ____ hundreds
____ tens
____ ones

**12.** 803 = ____ hundreds
____ tens
____ ones

**13.** 20 = ____ tens
____ ones

**14.** 116 = ____ hundreds
____ tens
____ ones

**15.** 200 = ____ hundreds
____ tens
____ ones

**Check your answers on page 152.**

# Forming Numbers from Place Values

Models can show how to add together place values to form whole numbers.

These models show the whole number 174.

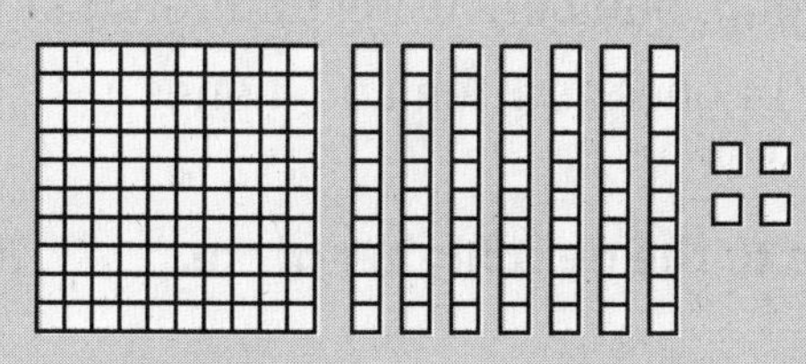

1 hundred + 7 tens + 4 ones = 100 + 70 + 4 = 174

What whole number do these models show?

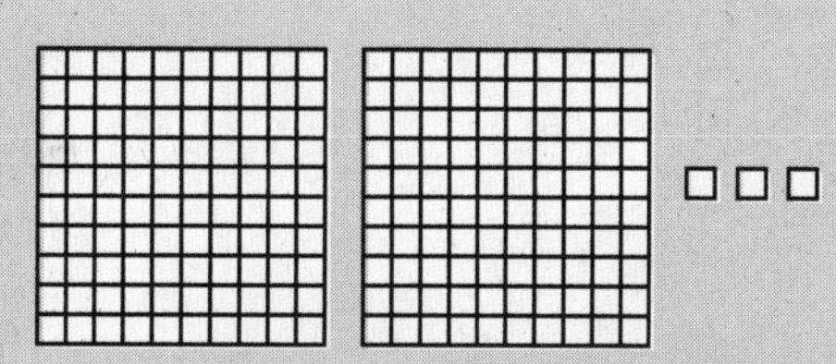

**2 hundreds + 0 tens + 3 ones = 200 + 3 = 203**
**There are 0 tens in the number, so write 0 in the tens place.**

**Write the numbers shown by these models.**

**1.**

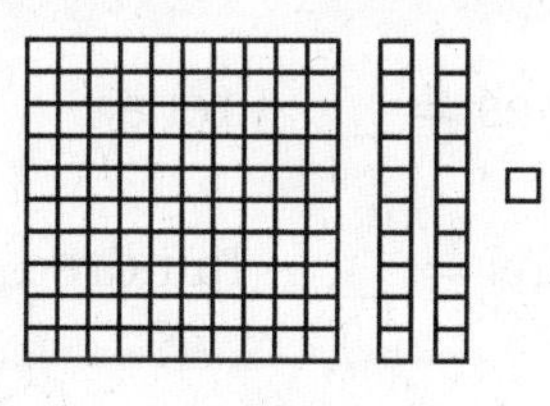

121

**2.**

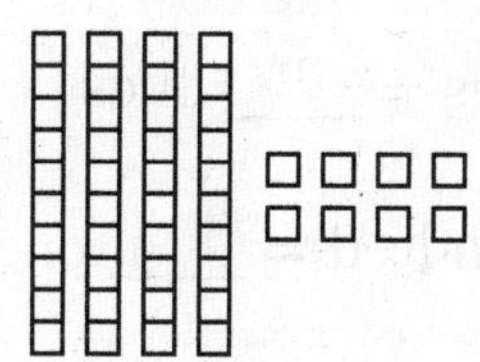

______

**3.**

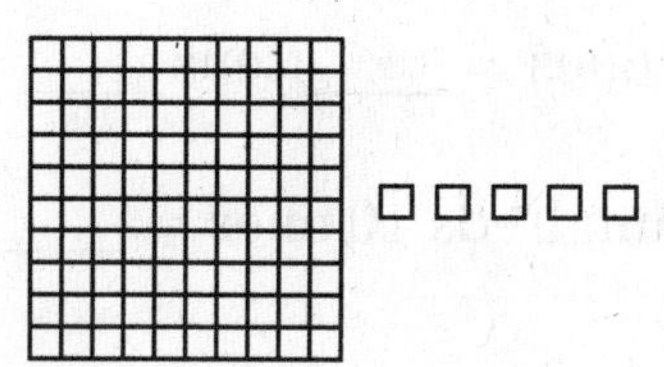

______

**4.**

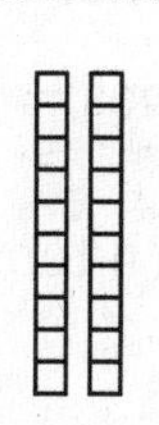

______

**Answer the word problems.**

**5.** At the post office Cletus asked for 346 stamps. He received 3 sheets of one hundred stamps, 4 strips of ten stamps, and 6 loose stamps. Did Cletus receive the correct number of stamps?

**6.** Michelle collected 7 sheets of one hundred bird stamps, 3 strips of ten deer stamps, and 5 loose stamps. How many stamps did she have in all?

**Check your answers on page 152.**

# Renaming Whole Numbers

To add or subtract whole numbers, sometimes you have to **rename**. Rename, or regroup, numbers using different place values. For example, you can rename 14 ones as 1 ten and 4 ones.

**Rename 10 ones to the next higher place.**

Rename 10 ones.

10 ones are renamed as 1 ten.

**10 ones = 1 ten**

**Rename 2 tens to the next lower place.**

Rename 2 tens.

2 tens are renamed as 20 ones.

**2 tens = 20 ones**

**Rename each of the following.**

1. 3 tens = 30 ones
2. 40 ones = _____ tens
3. 1 hundred = _____ ones
4. 10 tens = _____ hundred
5. 100 ones = _____ hundred
6. 1 hundred = _____ tens
7. 2 tens 4 ones = _____ ones
8. 39 ones = _____ tens _____ ones
9. 1 hundred 4 tens 6 ones = _____ ones
10. 2 hundreds 10 ones = _____ tens

**Answer the word problems.**

11. You have 3 notepads of 100 sheets and 4 pads of 10 sheets. How many sheets do you have in all?

12. You have 2 boxes of 100 paper clips and 4 chains of 10 paper clips each. How many paper clips do you have all together?

**Check your answers on page 152.**

# Place Value to Thousands

A place value chart can also show thousands, ten thousands, and hundred thousands. Commas are used to separate numbers into groups of three digits. This makes numbers easier to read.

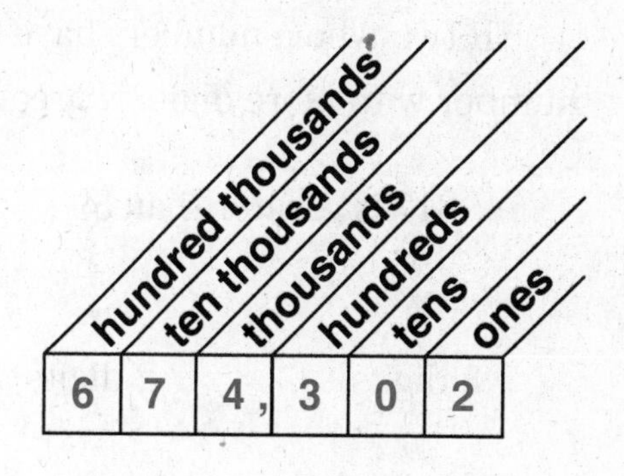


Look at the number 674,302 in the chart. The digit 2 is in the ones place, the digit 0 is in the tens place, the digit 3 is in the hundreds place, the digit 4 is in the thousands place, the digit 7 is in the ten-thousands place, and the digit 6 is in the hundred-thousands place.

**Write each number in the chart. Next write each digit in the correct number group below. Then write the value of each digit.**

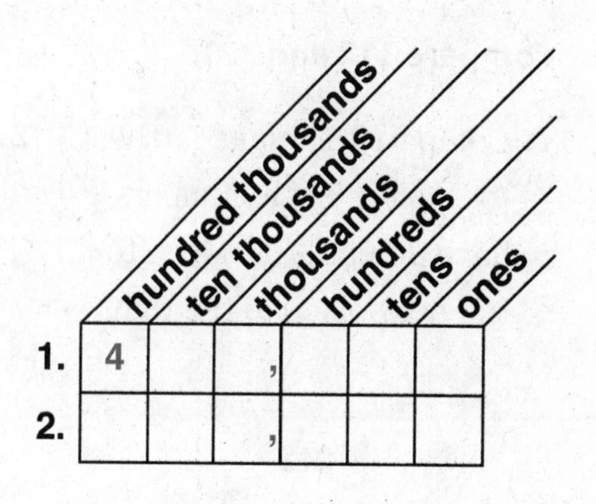


**1.** 459,328

__4__ hundred thousands = __400,000__

____ ten thousands = ________

____ thousands = ________

____ hundreds = ________

____ tens = ________

____ ones = ________

**2.** 47,006

____ ten thousands = ________

____ thousands = ________

____ hundreds = ________

____ tens = ________

____ ones = ________

**Write the value of the underlined digit in each number.**

**3.** 3,<u>4</u>25 __400__ **4.** 1<u>6</u>9,450 ________ **5.** 2<u>7</u>,998 ________

**6.** 144,3<u>0</u>8 ________ **7.** 982,<u>0</u>07 ________ **8.** <u>4</u>80,350 ________

Check your answers on page 152.

# Comparing Whole Numbers

Place value can be used to compare any two whole numbers.

If two whole numbers have different numbers of digits, the number with more digits is greater.

| 317 is greater than 89 | | 1,292 is greater than 689 | |
|---|---|---|---|
| ↓ | ↓ | ↓ | ↓ |
| 3 digits | 2 digits | 4 digits | 3 digits |

If two whole numbers have the same number of digits, compare each digit, beginning with the digits farthest to the left. Use the symbols > and < to compare. The symbol > means *greater than*. The symbol < means *less than*.

**Compare 312 and 321.**

**1.** Compare the digits farthest to the left. Both numbers have 3 hundreds. Move to the next place to the right.

312
321

**2.** Compare the numbers in the tens place. 321 has 2 tens. 312 has only 1 ten.

312
321

**3.** Use the symbols > and < to compare the numbers.

**321 is *greater than* 312.**
**$321 > 312$**

**312 is *less than* 321.**
**$312 < 321$**

**Compare each set of numbers. Write > for *greater than* or < for *less than*.**

| | | | | | | | |
|---|---|---|---|---|---|---|---|
| **1.** | 542 [ > ] 524 | **2.** | 47 [ ] 463 | **3.** | 4 [ ] 9 | **4.** | 83 [ ] 80 |
| **5.** | 212 [ ] 221 | **6.** | 630 [ ] 603 | **7.** | 8 [ ] 5 | **8.** | 34 [ ] 43 |
| **9.** | 502 [ ] 520 | **10.** | 144 [ ] 195 | **11.** | 2,276 [ ] 133 | **12.** | 61 [ ] 580 |
| **13.** | 57 [ ] 75 | **14.** | 330 [ ] 4,334 | **15.** | 988 [ ] 899 | **16.** | 403 [ ] 304 |
| **17.** | 6,295 [ ] 4,859 | **18.** | 432 [ ] 423 | **19.** | 818 [ ] 811 | **20.** | 8,349 [ ] 8,344 |
| **21.** | 789 [ ] 879 | **22.** | 1,750 [ ] 1,705 | **23.** | 389 [ ] 390 | **24.** | 201 [ ] 210 |

**Check your answers on page 152.**

# Rounding

Sometimes you do not need to give an exact answer to a math question. Suppose you were asked how much time you spend on the telephone each day. You might answer, "About twenty minutes." This answer describes *about* how much time, not *exactly* how much. When an exact answer is not necessary, you can **round** numbers. You can use a number line to round numbers.

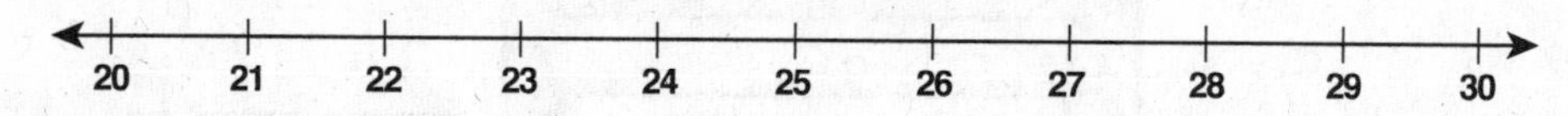


**Round 22 to the nearest ten.**

Find 22 on the number line. Then find the nearest tens numbers (20 and 30). Since 22 is closer to 20, 22 rounds down to 20.

**Round 27 to the nearest ten.**

Find 27 on the number line. Then find the nearest tens numbers (20 and 30). Since 27 is closer to 30, 27 rounds up to 30.

**Round 25 to the nearest ten.**

Find 25 on the number line. Then find the nearest tens numbers (20 and 30). Since 25 is exactly halfway between 20 and 30, round up to 30.

**Round each number to the nearest ten.**

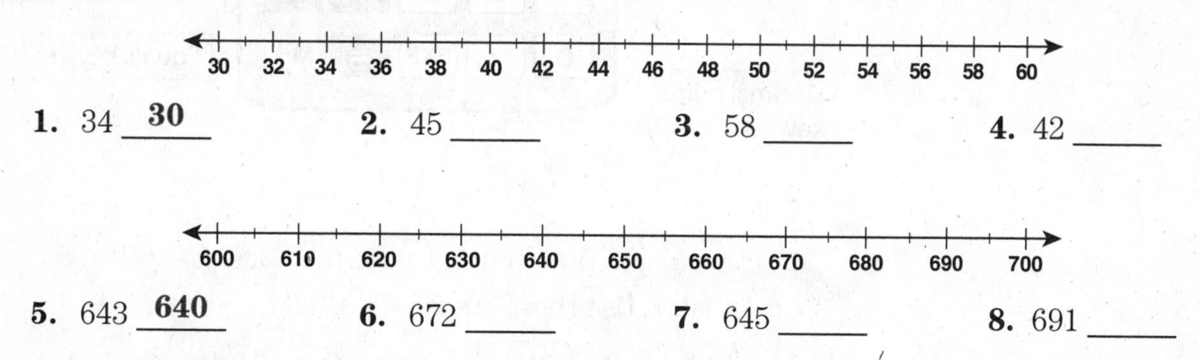


**1.** 34 **30**  **2.** 45 ______  **3.** 58 ______  **4.** 42 ______

**5.** 643 **640**  **6.** 672 ______  **7.** 645 ______  **8.** 691 ______

**Read each question. Would you give an exact answer or a rounded answer?**

**9.** How many children are in your family?

**10.** About how many hours do you sleep each week?

**Check your answers on page 153.**

# Using a Calculator

A calculator is a tool to help you solve problems and check your answers. Study the picture of the calculator used on the GED Math Test.

**The CASIO *fx-260SOLAR* Calculator**

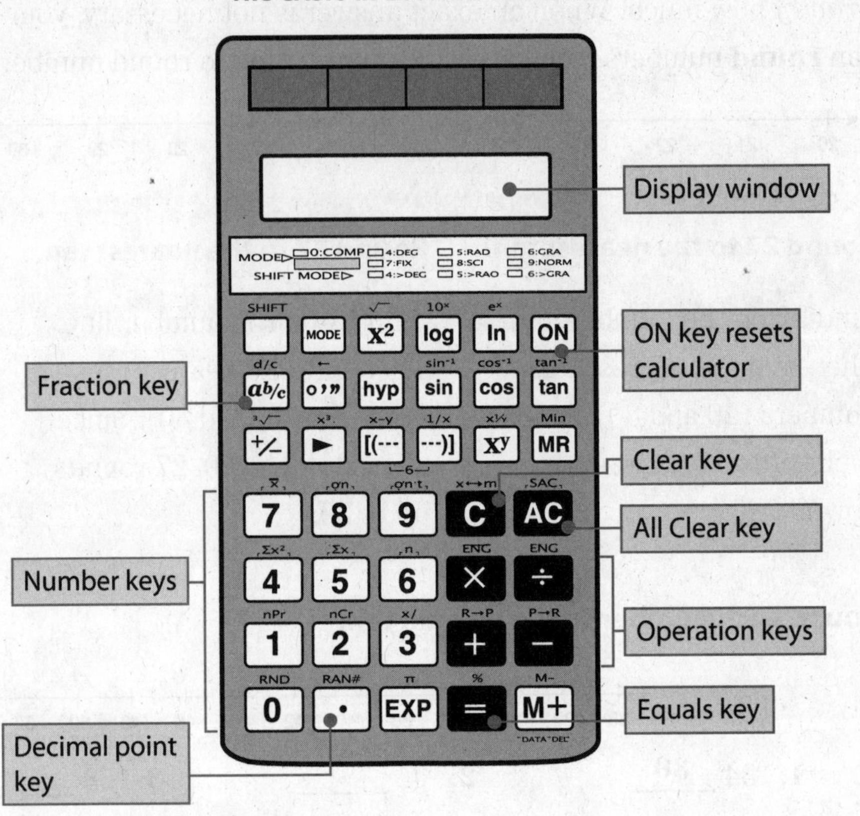


**Strategy** Learn how to use the Casio *fx-260* calculator. Use these steps.

1. Press the ON key to turn the calculator on.
2. Enter the number 7750 using the number keys.

3. Read the number on the display window. The calculator display does not show commas.

   7750.

4. Clear the display by pressing the AC key.

**Exercise 1: Enter each number below in your calculator. Be sure to clear your display with the AC key before each entry.**

**1.** 29

**2.** 307

**3.** 6,280

**4.** 60,280

**5.** 889,002

**6.** 223,468

**Exercise 2: Match each number with the way it would look on the calculator display.**

| | | | |
|---|---|---|---|
| c | **7.** 114 | **a.** | 140. |
| ____ | **8.** 11,140 | **b.** | 1140. |
| ____ | **9.** 14 | **c.** | 114. |
| ____ | **10.** 1,140 | **d.** | 14. |
| ____ | **11.** 140 | **e.** | 11140. |

**Exercise 3: Read the calculator displays below. Write the number shown on each display. Add commas where necessary.**

**12.** 98077. ______98,077______

**13.** 76222. ____________

**14.** 4600. ____________

**15.** 23756. ____________

**16.** 39641. ____________

**17.** 1842. ____________

**18.** 594437. ____________

**19.** 9046. ____________

Check your answers on page 153.

## Restating the Question

Before you can solve a word problem, you need to understand what the question is asking. To do this, find the question and look for important words and facts. Try to restate the question in your own words.

**Strategy** Try the strategy on the example below. Use these steps.

**Step 1** Read the word problem.

**Step 2** Find the question and restate it in your own words.

**Step 3** Solve the word problem.

**EXAMPLE**

**There are 4 packages of 10 juice boxes on the shelf. What is the total number of juice boxes on the shelf?**

(1) 14
(2) 40
(3) 6
(4) 44

In Step 1 you read the problem. In Step 2 you found the question and restated it in your own words. Your restated question could be something like: How many juice boxes are there in all? In Step 3 you solved the word problem. There are 4 packages of 10 juice boxes, or 4 sets of 10. Therefore, there are 40 juice boxes in all. The correct answer is (2).

## Practice

**Practice the strategy. Use the steps you learned. Solve the problem.**
**Circle the number of the correct answer.**

1. Ben used a counter to track the number of people at the water park. What is the value of the digit 4 in this number?

(1) ones
(2) tens
(3) hundreds
(4) thousands

2. Wendy makes 4 groups of 100 paint cans and 5 groups of 10 paint cans. How many cans of paint is this in all?

(1) 450
(2) 145
(3) 405
(4) 415

3. Liza cashes a check. She gets three 10-dollar bills and six 1-dollar bills. How much money does Liza get in all?

(1) 16
(2) 19
(3) 36
(4) 63

4. A shoe store is having a sneaker sale. Rounded to the nearest ten, about how much does a pair of sneakers cost?

(1) $40
(2) $30
(3) $20
(4) $10

5. It took 162 gallons to fill an oil tank. The digit 1 in this number is in what place?

(1) ones
(2) tens
(3) hundreds
(4) thousands

**Check your answers on page 153.**

# Unit 1 Wrap-up

Below are examples of the skills you learned in this unit. Read the examples and do the problems. Then check your answers.

**Examples.**

1. What is the place value of each digit in 68?

   **6 tens = 60**

   **8 ones = 8**

2. Write the place value of the underlined digit in 73.

   **7 tens**

3. 274 = **2 hundreds**

   **7 tens**

   **4 ones**

4. Write the whole number for 8 hundreds 3 tens 6 ones.

   **836**

5. Write the value of the underlined digit in 928,347.

   **20,000**

6. Compare 26 and 19.

   **26 > 19**

7. Round 59 to the nearest 10.

   **60**

**Problems.**

1. What is the place value of each digit in 902?

2. Write the place value of the underlined digit in 584.

3. 658 = ______ hundreds

   ______ tens

   ______ ones

4. Write the whole number for 5 hundreds 7 tens 2 ones.

5. Write the value of the underlined digit in 617,342.

6. Compare 37 and 73.

7. Round 125 to the nearest ten.

**Check your answers on page 153.**

## Unit 1 Practice

**Answer these questions.**

**1.** What is the place value of 8 in 87? _____ **2.** What is the place value of 4 in 24,308? _____

**Write each of these place values as a number.**

**3.** 4 tens 1 one _____ **4.** 9 hundreds 2 tens _____

**5.** 7 hundreds 6 tens 5 ones _____ **6.** 5 ones _____

**Rename each of the following numbers.**

**7.** 30 hundreds = _____ thousands **8.** 1 hundred = _____ ones

**9.** 20 tens = _____ hundreds **10.** 50 ones = _____ tens

**Compare each set of numbers. Write > or <.**

**11.** 3 ☐ 4 **12.** 76 ☐ 67 **13.** 220 ☐ 202 **14.** 53 ☐ 51

**15.** 437 ☐ 473 **16.** 330 ☐ 33 **17.** 41 ☐ 40 **18.** 78 ☐ 87

**Round each number to the nearest ten.**

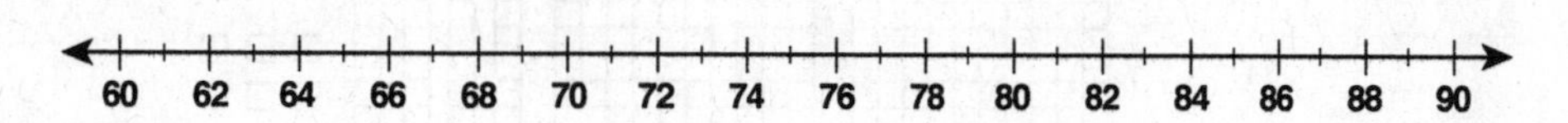


**19.** 84 _____ **20.** 67 _____ **21.** 75 _____ **22.** 88 _____

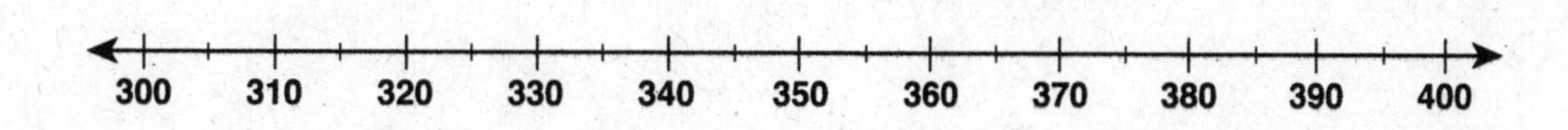


**23.** 315 _____ **24.** 362 _____ **25.** 306 _____ **26.** 329 _____

**Answer the word problems.**

**27.** Yesterday Luis delivered 32 newspapers and Sharise delivered 45. Who delivered more newspapers?

**28.** Mark works part-time 28 hours each week. Rounded to the nearest ten, about how many hours does Mark work each week?

**Check your answers on page 154.**

**Read each problem carefully. Circle the number of the correct answer.**

**1.** A company sent out 36,845 catalogs. What is the value of the digit 3 in this number?

(1) 300
(2) 3,000
(3) 30,000
(4) 300,000

**2.** A cereal box has 515 grams of cereal. Rounded to the nearest ten, how many grams of cereal are in the cereal box?

(1) 500
(2) 510
(3) 520
(4) 600

**3.** What number does this model show?

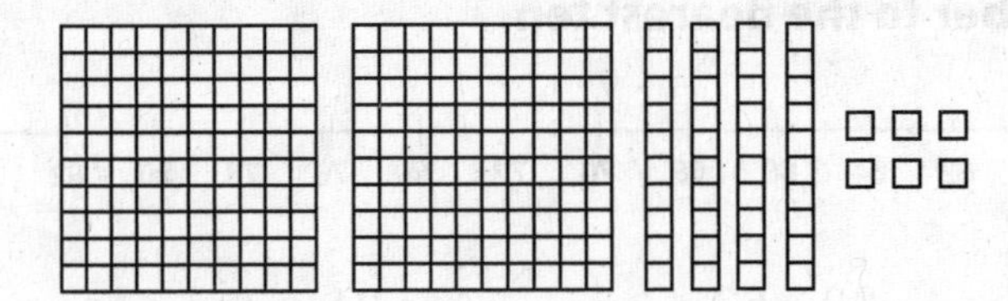

(1) 240
(2) 624
(3) 462
(4) 246

**4.** Eric can choose from 142 television channels. What is the place name for the digit 4 in this number?

(1) ones
(2) tens
(3) hundreds
(4) thousands

5. Marci ordered 2 packs of 100 red folders. She also ordered 5 packs of 10 blue folders. How many folders did she order in all?

(1) 250
(2) 200
(3) 150
(4) 70

6. In section A of the garage, there are 220 parking spaces. What is this number renamed in tens?

(1) 2 tens
(2) 20 tens
(3) 22 tens
(4) 220 tens

7. Look at this ticket. What is the value of the digit 7 in the ticket number?

(1) 7
(2) 70
(3) 700
(4) 7,000

8. Last night Emma read 67 pages in her book. Rounded to the nearest ten, about how many pages did she read?

(1) 100
(2) 70
(3) 65
(4) 60

9. Gabby's camera cost $247. Rounded to the nearest ten, how much did Gabby pay?

(1) $250
(2) $245
(3) $240
(4) $255

10. There were 38,942 fans at a football game. In what place is the digit 3 in this number?

(1) hundred thousands
(2) ten thousands
(3) thousands
(4) hundreds

**Check your answers on page 154.**

## Unit 1 Skill Check-Up Chart

Check your answers. In the first column, circle the numbers of any questions that you missed. Then look across the rows to see the skills you need to review and the pages where you can find each skill.

| Question | Skill | Page |
|---|---|---|
| 1, 4, 7 | Whole Number Place Names and Values | 10–11 |
| 2, 8, 9 | Rounding | 17 |
| 3 | Finding Place Values for Numbers | 12 |
| 5 | Forming Numbers from Place Values | 13 |
| 6 | Renaming Whole Numbers | 14 |
| 10 | Place Value to Thousands | 15 |

# Unit 2 Adding and Subtracting Whole Numbers

lead digit

amount

parentheses

estimate

**In this unit you will learn about**

- addition and subtraction facts
- addition and subtraction with renaming
- the order of operations
- rounding and estimating
- using a calculator to add and subtract
- addition and subtraction word problems

Addition and subtraction are operations you perform on numbers. Adding is finding the total of two or more numbers.

You add numbers every day. You might add to find how much money is in your pocket or how many eggs you have left. Write an example of numbers you might add, such as the number of minutes it took you to get ready this morning.

_______________________________________________

Subtracting is finding the difference between numbers. When you subtract, you take one amount away from another amount.

Write an example of how you have used subtraction, such as figuring out how much money you'll have left after buying lunch.

_______________________________________________

# Addition Facts

This table shows basic addition facts. Addition facts are used whenever you add numbers.

What is 2 + 3? Find the row that begins with 2 and the column that begins with 3. (Rows go across and columns go up and down.) The answer is the number (5) in the box where the row and the column meet.

| + | 0 | 1 | 2 | 3 | 4 | 5 | 6 | 7 | 8 | 9 |
|---|---|---|---|---|---|---|---|---|---|---|
| 0 | 0 | 1 | 2 | 3 | 4 | 5 | | | | |
| 1 | 1 | 2 | 3 | 4 | 5 | | | | | |
| 2 | 2 | 3 | 4 | 5 | | | | | | |
| 3 | 3 | 4 | 5 | | | | | | | |
| 4 | 4 | 5 | | | | | | | | |
| 5 | 5 | | | | | | | | | |
| 6 | | | | | | | | | | |
| 7 | | | | | | | | | | |
| 8 | | | | | | | | | | |
| 9 | | | | | | | | | | |

**Complete this addition facts table. Look for patterns as you work.**

**Complete the following addition facts.**

**1.** $\begin{array}{r} 4 \\ +6 \\ \hline \mathbf{10} \end{array}$ **2.** $\begin{array}{r} 7 \\ +2 \\ \hline \end{array}$ **3.** $\begin{array}{r} 5 \\ +5 \\ \hline \end{array}$ **4.** $\begin{array}{r} 2 \\ +3 \\ \hline \end{array}$ **5.** $\begin{array}{r} 5 \\ +6 \\ \hline \end{array}$

**6.** $\begin{array}{r} 1 \\ +8 \\ \hline \end{array}$ **7.** $\begin{array}{r} 9 \\ +7 \\ \hline \end{array}$ **8.** $\begin{array}{r} 2 \\ +2 \\ \hline \end{array}$ **9.** $\begin{array}{r} 6 \\ +8 \\ \hline \end{array}$ **10.** $\begin{array}{r} 9 \\ +9 \\ \hline \end{array}$

**11.** 8 + 9 = **12.** 7 + 6 = **13.** 9 + 3 = **14.** 5 + 9 =

**15.** 2 + 0 = **16.** 6 + 6 = **17.** 8 + 4 = **18.** 3 + 7 =

**Solve each problem.**

**19.** Nine plus six equals what number?

**9 + 6 = 15**

**20.** Add four more to seven.

______________________

**21.** What is six and four more?

______________________

**22.** Add five and two.

______________________

**Check your answers on page 154.**

# Adding Larger Numbers

You can use the addition facts to add larger numbers. Add only the digits that have the same place value.

**Add:** $\begin{array}{r} 841 \\ +\,3{,}026 \\ \hline \end{array}$

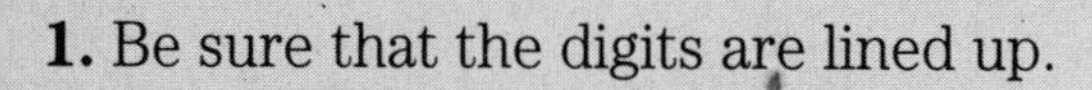

**1.** Be sure that the digits are lined up.

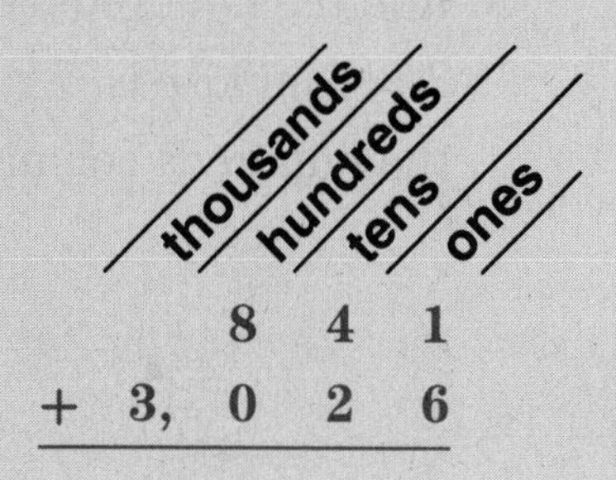


**2.** Add each column. Start with the digits in the ones place.

$$\begin{array}{r} 841 \\ +\,3{,}026 \\ \hline 7 \end{array} \qquad \begin{array}{r} 841 \\ +\,3{,}026 \\ \hline 67 \end{array} \qquad \begin{array}{r} 841 \\ +\,3{,}026 \\ \hline 867 \end{array} \qquad \begin{array}{r} 841 \\ +\,3{,}026 \\ \hline 3{,}867 \end{array}$$

↑ **ones** ↑ **tens** ↑ **hundreds** ↑ **thousands**

**Add.**

**1.** $\begin{array}{r} 13 \\ +\,85 \\ \hline 98 \end{array}$ **2.** $\begin{array}{r} 92 \\ +\,7 \\ \hline \end{array}$ **3.** $\begin{array}{r} 4 \\ +\,62 \\ \hline \end{array}$ **4.** $\begin{array}{r} 20 \\ +\,35 \\ \hline \end{array}$

**5.** $\begin{array}{r} 17 \\ +\,40 \\ \hline \end{array}$ **6.** $\begin{array}{r} 321 \\ +\,48 \\ \hline \end{array}$ **7.** $\begin{array}{r} 42 \\ +\,557 \\ \hline \end{array}$ **8.** $\begin{array}{r} 315 \\ +\,663 \\ \hline \end{array}$

**9.** $\begin{array}{r} 252 \\ +\,115 \\ \hline \end{array}$ **10.** $\begin{array}{r} 740 \\ +\,126 \\ \hline \end{array}$ **11.** $\begin{array}{r} 500 \\ +\,35 \\ \hline \end{array}$ **12.** $\begin{array}{r} 54 \\ +\,702 \\ \hline \end{array}$

**13.** $\begin{array}{r} 58 \\ +\,430 \\ \hline \end{array}$ **14.** $\begin{array}{r} 240 \\ +\,603 \\ \hline \end{array}$ **15.** $\begin{array}{r} 15 \\ +\,670 \\ \hline \end{array}$ **16.** $\begin{array}{r} 1{,}342 \\ +\,534 \\ \hline \end{array}$

**17.** $\begin{array}{r} 213 \\ +\,4{,}665 \\ \hline \end{array}$ **18.** $\begin{array}{r} 5{,}134 \\ +\,2{,}632 \\ \hline \end{array}$ **19.** $\begin{array}{r} 7{,}244 \\ +\,1{,}513 \\ \hline \end{array}$ **20.** $\begin{array}{r} 4{,}000 \\ +\,53 \\ \hline \end{array}$

**21.** $\begin{array}{r} 250 \\ +\,5{,}300 \\ \hline \end{array}$ **22.** $\begin{array}{r} 1{,}086 \\ +\,4{,}002 \\ \hline \end{array}$ **23.** $\begin{array}{r} 2{,}089 \\ +\,6{,}100 \\ \hline \end{array}$

**Check your answers on page 155.**

# Adding with Renaming

When you add the digits in a column, the sum can sometimes be 10 or more. When this happens, rename by carrying over to the next column to the left. You may need to rename more than once.

**Add:** 454 + 276

1. Add the ones. $4 + 6 = 10$ ones. Write 0 in the ones column. Carry 1 ten to the next column.

1 ten →

$$\begin{array}{r} {}^{1}\phantom{00} \\ 454 \\ +\,276 \\ \hline \mathbf{0} \end{array}$$

↑ **ones**

2. Add the tens. $1 + 5 + 7 =$ 13 tens. Write 3 in the tens column. Carry 1 hundred.

1 hundred →

$$\begin{array}{r} {}^{1\,1}\phantom{0} \\ 454 \\ +\,276 \\ \hline \mathbf{30} \end{array}$$

↑ **tens**

3. Add the hundreds. $1 + 4 + 2 = 7$ hundreds. Write 7 in the hundreds column.

$$\begin{array}{r} {}^{1\,1}\phantom{0} \\ 454 \\ +\,276 \\ \hline \mathbf{730} \end{array}$$

↑ **hundreds**

**Add.**

1. $\begin{array}{r} {}^{1}\phantom{0} \\ 57 \\ +\,94 \\ \hline \mathbf{151} \end{array}$

2. $\begin{array}{r} 68 \\ +\,6 \\ \hline \end{array}$

3. $\begin{array}{r} 2 \\ +\,59 \\ \hline \end{array}$

4. $\begin{array}{r} 44 \\ +\,36 \\ \hline \end{array}$

5. $\begin{array}{r} 78 \\ +\,49 \\ \hline \end{array}$

6. $\begin{array}{r} 15 \\ +\,85 \\ \hline \end{array}$

7. $\begin{array}{r} 19 \\ +\,4 \\ \hline \end{array}$

8. $\begin{array}{r} 3 \\ +\,77 \\ \hline \end{array}$

9. $\begin{array}{r} 34 \\ +\,76 \\ \hline \end{array}$

10. $\begin{array}{r} 43 \\ +\,8 \\ \hline \end{array}$

11. $\begin{array}{r} 547 \\ +\,67 \\ \hline \end{array}$

12. $\begin{array}{r} 31 \\ +\,693 \\ \hline \end{array}$

13. $\begin{array}{r} 117 \\ +\,914 \\ \hline \end{array}$

14. $\begin{array}{r} 454 \\ +\,709 \\ \hline \end{array}$

15. $\begin{array}{r} 236 \\ +\,481 \\ \hline \end{array}$

16. $\begin{array}{r} 1{,}215 \\ +\,883 \\ \hline \end{array}$

17. $\begin{array}{r} 4{,}070 \\ +\,4{,}035 \\ \hline \end{array}$

18. $\begin{array}{r} 1{,}968 \\ +\,2{,}732 \\ \hline \end{array}$

19. $\begin{array}{r} 5{,}140 \\ +\,6{,}380 \\ \hline \end{array}$

**Check your answers on page 155.**

# Adding Three or More Numbers

To add three or more numbers, add the numbers one column at a time, working from right to left. Add the digits in pairs.

**Add:**
$$\begin{array}{r} 58 \\ 16 \\ +\ 32 \\ \hline \end{array}$$

1. Start with groups of digits in the ones column. Add the digits in pairs.

$$\begin{array}{r} \scriptstyle 1 \\ 58 \\ 16 \\ +\ 32 \\ \hline 6 \end{array} \qquad \begin{array}{l} 8 + 6 = 14 \\ 14 + 2 = 16 \end{array}$$

2. Add groups of digits in the tens column.

$$\begin{array}{l} 1 + 5 = 6 \\ 6 + 1 = 7 \\ 7 + 3 = 10 \end{array} \qquad \begin{array}{r} \scriptstyle 1 \\ 58 \\ 16 \\ +\ 32 \\ \hline 106 \end{array}$$

**Add.**

1. $\begin{array}{r} \scriptstyle 2 \\ 25 \\ 19 \\ +\ 6 \\ \hline 50 \end{array}$
2. $\begin{array}{r} 5 \\ 2 \\ +\ 6 \\ \hline \end{array}$
3. $\begin{array}{r} 53 \\ 38 \\ +\ 20 \\ \hline \end{array}$
4. $\begin{array}{r} 64 \\ 520 \\ +\ 50 \\ \hline \end{array}$
5. $\begin{array}{r} 24 \\ 279 \\ +\ 50 \\ \hline \end{array}$
6. $\begin{array}{r} 201 \\ 383 \\ 10 \\ +\ 301 \\ \hline \end{array}$
7. $\begin{array}{r} 212 \\ 311 \\ 203 \\ +\ 62 \\ \hline \end{array}$
8. $\begin{array}{r} 340 \\ 23 \\ 5 \\ +\ 219 \\ \hline \end{array}$
9. $\begin{array}{r} 217 \\ 224 \\ 70 \\ +\ 321 \\ \hline \end{array}$
10. $\begin{array}{r} 660 \\ 61 \\ 382 \\ +\ 17 \\ \hline \end{array}$

**Answer the word problems.**

11. Harriet's furnace used 38 gallons of fuel oil in December, 75 gallons in January, and 67 gallons in February. How many gallons of fuel oil did the furnace use in all?

12. By last Wednesday, Tara had written 192 tickets. On Thursday, she wrote 67 tickets, and on Friday she wrote 59 tickets. What was her weekly total?

**Check your answers on page 155.**

## Using a Calculator to Add

A calculator can help you add numbers. Always clear your calculator with the AC key before you begin. Check each number after you enter it.

**Strategy** Learn how to add numbers on the Casio *fx-260* calculator. Follow these steps.

**Add:** 78 + 105

| | Press the key. | Read the display. |
|---|---|---|
| **1.** Clear the calculator. | → AC | → 0. |
| **2.** Enter the first number. | → 7 8 | → 78. |
| **3.** Press the add key. | → + | → 78. |
| **4.** Enter the next number. | → 1 0 5 | → 105. |
| **5.** Press the equals key. | → = | → 183. |

**Exercise 1: Enter the numbers and symbols on your calculator. Write what the display shows.**

1. AC 5 6 ______
2. + ______
3. 4 8 ______
4. + ______
5. 1 8 9 ______
6. + ______
7. 5 5 ______
8. = ______
9. AC 2 3 7 ______
10. + ______
11. 6 6 3 ______
12. + ______
13. 1 1 4 5 ______
14. + ______
15. 5 9 0 ______
16. = ______

**Exercise 2: Use a calculator to add.**

**17.** $51 + 28$

**18.** $31 + 16$

**19.** $45 + 82$

**20.** $55 + 74$

**21.** $104 + 580$

**22.** $621 + 414$

**23.** $702 + 979$

**24.** $318 + 710$

**25.** $81 + 17 + 59$

**26.** $223 + 165 + 542$

**27.** $1{,}038 + 971 + 2{,}256$

**28.** $166 + 858 + 3{,}077$

**Exercise 3: Use a calculator to solve.**

**29.** This month Riko's cable television bill was $47, his telephone bill was $68, and his electric bill was $99. What is the total amount of these bills?

**30.** Lee drove 318 miles. The next day she drove 226 miles. What is the total number of miles Lee drove on those two days?

**31.** Last year Grace worked 223 days. This year she worked 238 days. In all, how many days did Grace work this year and last year?

**32.** Ramon took some items to the recycling center. He took 59 bottles and 76 cans. How many items is this in all?

**Check your answers on page 155.**

# Subtraction Facts

Subtraction is the opposite of addition. The addition facts table can help you solve subtraction problems. To find 14 – 8, find the smaller number (8) in the first column of the addition table. Next, move to the right to find the larger number (14). Finally move up to find the answer (6) in the first row.

| – | 0 | 1 | 2 | 3 | 4 | 5 | 6 | 7 | 8 | 9 |
|---|---|---|---|---|---|---|---|---|---|---|
| 0 | 0 | 1 | 2 | 3 | 4 | 5 | 6 | 7 | 8 | 9 |
| 1 | 1 | 2 | 3 | 4 | 5 | 6 | 7 | 8 | 9 | 10 |
| 2 | 2 | 3 | 4 | 5 | 6 | 7 | 8 | 9 | 10 | 11 |
| 3 | 3 | 4 | 5 | 6 | 7 | 8 | 9 | 10 | 11 | 12 |
| 4 | 4 | 5 | 6 | 7 | 8 | 9 | 10 | 11 | 12 | 13 |
| 5 | 5 | 6 | 7 | 8 | 9 | 10 | 11 | 12 | 13 | 14 |
| 6 | 6 | 7 | 8 | 9 | 10 | 11 | 12 | 13 | 14 | 15 |
| 7 | 7 | 8 | 9 | 10 | 11 | 12 | 13 | 14 | 15 | 16 |
| 8 | 8 | 9 | 10 | 11 | 12 | 13 | 14 | 15 | 16 | 17 |
| 9 | 9 | 10 | 11 | 12 | 13 | 14 | 15 | 16 | 17 | 18 |

**Use the table to complete the following subtraction facts.**

**1.** $\begin{array}{r} 11 \\ -\ 5 \\ \hline \mathbf{6} \end{array}$ **2.** $\begin{array}{r} 6 \\ -2 \\ \hline \end{array}$ **3.** $\begin{array}{r} 10 \\ -\ 7 \\ \hline \end{array}$ **4.** $\begin{array}{r} 11 \\ -\ 3 \\ \hline \end{array}$ **5.** $\begin{array}{r} 16 \\ -\ 9 \\ \hline \end{array}$

**6.** $\begin{array}{r} 9 \\ -4 \\ \hline \end{array}$ **7.** $\begin{array}{r} 13 \\ -\ 6 \\ \hline \end{array}$ **8.** $\begin{array}{r} 9 \\ -0 \\ \hline \end{array}$ **9.** $\begin{array}{r} 8 \\ -8 \\ \hline \end{array}$ **10.** $\begin{array}{r} 7 \\ -5 \\ \hline \end{array}$

**11.** 11 – 2 = **12.** 13 – 4 = **13.** 12 – 6 = **14.** 15 – 8 =

**Solve each problem.**

**15.** Ten minus six equals what number?

**10 – 6 = 4**

**16.** What is eight less than eleven?

______________________

**17.** Subtract five from nine.

______________________

**18.** Find the difference between thirteen and six.

______________________

**Check your answers on page 156.**

# Subtracting Larger Numbers

You can use the subtraction facts to subtract larger numbers. Subtract only the digits that have the same place value.

**Subtract:** $\begin{array}{r} 237 \\ -\,125 \\ \hline \end{array}$

1. Be sure to line up the digits.

$$\begin{array}{r} 237 \\ -\,125 \\ \hline \end{array}$$

2. Subtract each column, starting with the digits in the ones place.

$$\begin{array}{r} 237 \\ -\,125 \\ \hline 2 \end{array} \qquad \begin{array}{r} 237 \\ -\,125 \\ \hline 12 \end{array} \qquad \begin{array}{r} 237 \\ -\,125 \\ \hline 112 \end{array}$$

3. Check by adding the answer (112) to the bottom number (125). The sum should be the same as the top number (237).

Check:

$$\begin{array}{r} 112 \\ +\,125 \\ \hline 237 \end{array}$$

**Subtract. Use addition to check your answers.**

**1.** $\begin{array}{r} 46 \\ -\,22 \\ \hline 24 \end{array} \quad \begin{array}{r} 24 \\ +\,22 \\ \hline 46 \end{array}$ **2.** $\begin{array}{r} 67 \\ -\,30 \\ \hline \end{array}$ **3.** $\begin{array}{r} 89 \\ -\,58 \\ \hline \end{array}$ **4.** $\begin{array}{r} 94 \\ -\,10 \\ \hline \end{array}$

**5.** $\begin{array}{r} 52 \\ -\,31 \\ \hline \end{array}$ **6.** $\begin{array}{r} 77 \\ -\,16 \\ \hline \end{array}$ **7.** $\begin{array}{r} 25 \\ -\,13 \\ \hline \end{array}$ **8.** $\begin{array}{r} 33 \\ -\,22 \\ \hline \end{array}$

**9.** $\begin{array}{r} 619 \\ -\quad 7 \\ \hline \end{array}$ **10.** $\begin{array}{r} 171 \\ -\ \ 51 \\ \hline \end{array}$ **11.** $\begin{array}{r} 438 \\ -\,126 \\ \hline \end{array}$ **12.** $\begin{array}{r} 794 \\ -\,503 \\ \hline \end{array}$

**13.** $\begin{array}{r} 3{,}487 \\ -\quad 340 \\ \hline \end{array}$ **14.** $\begin{array}{r} 1{,}195 \\ -\quad\ 72 \\ \hline \end{array}$ **15.** $\begin{array}{r} 4{,}864 \\ -\,1{,}602 \\ \hline \end{array}$ **16.** $\begin{array}{r} 7{,}637 \\ -\,3{,}214 \\ \hline \end{array}$

**Check your answers on page 156.**

# Subtracting with Renaming

In some problems, the digit you are subtracting from is too small. You can borrow from the next column to the left. Rename 1 ten as 10 ones, or 1 hundred as 10 tens.

**Subtract:** $\begin{array}{r} 256 \\ -\ 187 \\ \hline \end{array}$

1. Start in the ones place. You can't subtract 7 from 6. Borrow 1 ten and rename it as 10 ones. Subtract the ones.

$$\begin{array}{r} {\scriptstyle 4\,16} \\ 2\not{5}\not{6} \\ -\ 187 \\ \hline 9 \end{array}$$

2. You can't subtract 8 from 4. Borrow 1 hundred and rename it as 10 tens. Subtract the tens.

$$\begin{array}{r} {\scriptstyle 14} \\ {\scriptstyle 1\,\not{4}\,16} \\ \not{2}\not{5}\not{6} \\ -\ 187 \\ \hline 69 \end{array}$$

3. Subtract the hundreds. $1 - 1 = 0$ hundreds. Check your answer.

Check:

$$\begin{array}{r} {\scriptstyle 14} \\ {\scriptstyle 1\,\not{4}\,16} \\ \not{2}\not{5}\not{6} \\ -\ 187 \\ \hline 69 \end{array} \qquad \begin{array}{r} {\scriptstyle 1\,1} \\ 69 \\ +\ 187 \\ \hline 256 \end{array}$$

**Subtract. Use addition to check your answers.**

1. $\begin{array}{r} {\scriptstyle 2\,15} \\ \not{3}\not{5} \\ -\ 17 \\ \hline 18 \end{array} \quad \begin{array}{r} {\scriptstyle 1} \\ 18 \\ +\ 17 \\ \hline 35 \end{array}$

2. $\begin{array}{r} 46 \\ -\ 27 \\ \hline \end{array}$

3. $\begin{array}{r} 73 \\ -\ 6 \\ \hline \end{array}$

4. $\begin{array}{r} 61 \\ -\ 33 \\ \hline \end{array}$

5. $\begin{array}{r} 24 \\ -\ 19 \\ \hline \end{array}$

6. $\begin{array}{r} 83 \\ -\ 58 \\ \hline \end{array}$

7. $\begin{array}{r} 52 \\ -\ 46 \\ \hline \end{array}$

8. $\begin{array}{r} 94 \\ -\ 8 \\ \hline \end{array}$

9. $\begin{array}{r} 675 \\ -\ 290 \\ \hline \end{array}$

10. $\begin{array}{r} 342 \\ -\ 235 \\ \hline \end{array}$

11. $\begin{array}{r} 865 \\ -\ 74 \\ \hline \end{array}$

12. $\begin{array}{r} 413 \\ -\ 207 \\ \hline \end{array}$

13. $\begin{array}{r} 2{,}274 \\ -\ 1{,}092 \\ \hline \end{array}$

14. $\begin{array}{r} 5{,}381 \\ -\ 3{,}824 \\ \hline \end{array}$

15. $\begin{array}{r} 8{,}416 \\ -\ 4{,}018 \\ \hline \end{array}$

16. $\begin{array}{r} 6{,}172 \\ -\ 831 \\ \hline \end{array}$

**Check your answers on page 156.**

# Renaming Zeros

Some subtraction problems may have one or more zeros in the top number. When this happens, you need to borrow across the zero or zeros.

**Subtract:** 702 − 176

1. To subtract the ones, borrow 1 hundred across the zero from the hundreds column. Rename 1 hundred as 10 tens.

```
 6 10
 7 0 2
−  176
```

2. Borrow 1 ten and rename as 10 ones. Now there are 6 hundreds, 9 tens, and 12 ones.

```
   9
 6 10 12
 7 0 2
−  176
```

3. Subtract.
12 − 6 = 6 ones
9 − 7 = 2 tens
6 − 1 = 5 hundreds
Check your answer.

```
   9
 6 10 12          Check:
 7 0 2             1 1
−  176             526
   526           + 176
                   702
```

**Subtract. Use addition to check your answers.**

1. 504 − 369 = 135 (worked: 5 → 4, 0 → 10 → 9, 4 → 14); Check: 135 + 369 = 504
2. 703 − 39
3. 510 − 7
4. 802 − 54
5. 500 − 133 = 367 (worked: 5 → 4, 0 → 10 → 9, 0 → 10); Check: 367 + 133 = 500
6. 400 − 75
7. 600 − 18
8. 300 − 221
9. 106 − 64
10. 705 − 329
11. 201 − 103
12. 906 − 87
13. 3,502 − 265
14. 4,002 − 1,073
15. 903 − 258
16. 7,304 − 760

Check your answers on page 157.

## Using a Calculator to Subtract

A calculator can help you subtract numbers. Always clear your calculator with the [AC] key before you begin. Check each number after you enter it.

**Strategy** Learn how to subtract numbers on the Casio *fx-260* calculator. Follow these steps.

**Subtract: 846 – 479 – 75**

| | Press the key. | Read the display. |
|---|---|---|
| **1.** Clear the calculator. | → [AC] | → 0. |
| **2.** Enter the larger number. | → [8] [4] [6] | → 846. |
| **3.** Press the subtract key. | → [–] | → 846. |
| **4.** Enter the next number. | → [4] [7] [9] | → 479. |
| **5.** Press the subtract key. | → [–] | → 367. |
| **6.** Enter the next number. | → [7] [5] | → 75. |
| **7.** Press the equals key. | → [=] | → 292. |

**Exercise 1: Enter the numbers and symbols on your calculator. Write what the display shows.**

**1.** [AC] [8] [4] ______

**2.** [–] ______

**3.** [3] [9] ______

**4.** [–] ______

**5.** [1] [9] ______

**6.** [=] ______

**7.** [AC] [1] [0] [7] [6] ______

**8.** [–] ______

**9.** [5] [3] [9] ______

**10.** [–] ______

**11.** [2] [5] [0] ______

**12.** [=] ______

**Exercise 2: Use a calculator to subtract.**

**13.** $85 - 51 =$ **14.** $\$91 - \$74 =$ **15.** $44 - 18 =$ **16.** $49 - 37 =$

**17.** $\$73 - \$24 =$ **18.** $42 - 15 =$ **19.** $96 - 59 =$

**20.** $\begin{array}{r} 487 \\ -\ 158 \\ \hline \end{array}$ **21.** $\begin{array}{r} 385 \\ -\ 166 \\ \hline \end{array}$ **22.** $\begin{array}{r} 631 \\ -\ 490 \\ \hline \end{array}$ **23.** $\begin{array}{r} \$303 \\ -\ 264 \\ \hline \end{array}$

**24.** $820 - 157 - 82 =$ **25.** $\$806 - \$293 =$ **26.** $1{,}762 - 311 - 800 =$

**Exercise 3: Use a calculator to solve.**

**27.** Rita wants to save $225 to buy a television set. She has already saved $160. How much more does Rita have to save?

**28.** A painter spent $45 on paint and $28 on supplies. How much more did the paint cost than the supplies?

**29.** A local firehouse received calls 317 days last year. On how many days were no calls received? (1 year = 365 days)

**30.** Stan has 72 stamps. He uses 18 stamps to send out some letters. How many stamps are left?

**Check your answers on page 157.**

# Lining up Numbers in Adding and Subtracting

Sometimes you may need to line up numbers to add. First put the digits in columns. When you need to carry, remember to write the number you carry above the column to the left.

**Add: 396 + 389**

**1.** Write the digits in columns so that the digits with the same place value are lined up.

$$\begin{array}{r} 396 \\ +\ 389 \\ \hline \end{array}$$

**2.** Add the digits in each column. Start with the ones place. Work from right to left.

$$\begin{array}{r} {\scriptstyle 1} \\ 396 \\ +\ 389 \\ \hline 5 \end{array} \qquad \begin{array}{r} {\scriptstyle 11} \\ 396 \\ +\ 389 \\ \hline 85 \end{array} \qquad \begin{array}{r} {\scriptstyle 11} \\ 396 \\ +\ 389 \\ \hline 785 \end{array}$$

**Add.**

**1.** 56 + 316 =

$$\begin{array}{r} {\scriptstyle 1} \\ 56 \\ +\ 316 \\ \hline 372 \end{array}$$

**2.** 37 + 88 =

**3.** 76 + 5 =

**4.** 35 + 19 =

**5.** 306 + 1,174 =

**6.** 3,607 + 453 =

**7.** 2,217 + 1,855 =

**8.** 9,528 + 4,490 =

**9.** 48 + 261 =

**10.** 13 + 77 =

**11.** 405 + 58 =

**12.** 23 + 4,081 =

**Answer the word problems.**

**13.** Trish spent $125 for a new winter coat, plus $28 for gloves and a scarf. How much did Trish spend in all?

**14.** Al held a 2-day yard sale. The first day Al made $135. The second day Al made $68. How much money did Al make all together?

**Check your answers on page 158**

Sometimes you may need to line up numbers to subtract. First, put the digits in columns. Make sure the larger number is on top. Remember to write the renamed numbers above the correct top numbers.

**Subtract: 267 – 53**

1. Write the digits in columns so that the digits with the same place value are lined up.

$$\begin{array}{r} 267 \\ -\ 53 \\ \hline \end{array}$$

2. Subtract the digits in each column, starting with the ones place.

$$\begin{array}{r} 267 \\ -\ 53 \\ \hline 4 \end{array} \qquad \begin{array}{r} 267 \\ -\ 53 \\ \hline 14 \end{array} \qquad \begin{array}{r} 267 \\ -\ 53 \\ \hline 214 \end{array} \qquad \begin{array}{r} 214 \\ +\ 53 \\ \hline 267 \end{array}$$

**Subtract. Use addition to check your answers.**

1. 48 – 6 =

$$\begin{array}{r} 48 \\ -\ 6 \\ \hline 42 \end{array} \qquad \begin{array}{r} 42 \\ +\ 6 \\ \hline 48 \end{array}$$

2. 56 – 23 =
3. 65 – 14 =
4. $76 – $24 =
5. 471 – 126 =
6. $817 – $174 =
7. 652 – 87 =
8. 801 – 233 =
9. 8,527 – 315 =
10. $4,340 – $220 =
11. $3,204 – $2,818 =
12. 7,003 – 4,625 =

**Answer the word problems.**

13. At a garage sale, a charity group earned $865 on Saturday and $520 on Sunday. How much more did the group earn on Saturday than on Sunday?

14. Clayton has traveled 325 miles so far. He must drive 638 miles in all. How many miles does Clayton have left to travel?

**Check your answers on page 158.**

# Addition and Subtraction: Two or More Steps

When a problem has more than two steps, work from left to right.

**Solve: 231 – 107 + 88 – 125**

1. Start at the left. Subtract the first two numbers 231 – 107. Line up the digits.

$$231 - 107 = \begin{array}{r} \overset{2\,11}{231} \\ -\,107 \\ \hline 124 \end{array}$$

2. Add 88 to the answer from Step 1.

$$\begin{array}{r} \overset{1\,1}{124} \\ +\,88 \\ \hline 212 \end{array}$$

3. Subtract 125 from the answer from Step 2.

$$\begin{array}{r} \overset{1\,10\,12}{212} \\ -\,125 \\ \hline 87 \end{array}$$

**Solve. Work from left to right.**

1. 19 + 27 – 13 =

$$\begin{array}{r} \overset{1}{19} \\ +\,27 \\ \hline 46 \end{array} \quad \begin{array}{r} 46 \\ -\,13 \\ \hline 33 \end{array}$$

2. 76 – 19 + 23 =
3. 56 + 12 – 41 =
4. 81 – 5 – 32 =
5. 302 – 118 + 65 =
6. \$58 + \$476 – \$109 =
7. \$689 – \$47 – \$124 =

**Use a calculator to solve. Work from left to right.**

8. \$36 + \$8 + \$16 =
9. 90 – 35 + 26 =
10. 13 + 53 – 28 =
11. 586 + 442 – 703 =
12. \$406 – \$168 + \$635 =
13. 335 + 921 – 274 =

**Answer the word problems.**

14. Greg bought a birthday cake for \$13 and balloons for \$4. How much change should he get from \$20?
15. A concert hall has 3,000 seats. Last night 1,287 adults and 365 children attended. How many seats in the hall were empty?

**Check your answers on page 158.**

# Order of Operations

When a problem has parentheses ( ), do the operation inside the parentheses first.

**Solve: (18 + 83) – (25 + 19)**

**1.** Add 18 + 83.

$$\begin{array}{r} {}^{1}\phantom{8} \\ 18 \\ +\,83 \\ \hline 101 \end{array}$$

**2.** Add 25 + 19.

$$\begin{array}{r} {}^{1}\phantom{5} \\ 25 \\ +\,19 \\ \hline 44 \end{array}$$

**3.** Subtract the answers from Steps 1 and 2.

$$\begin{array}{r} {}^{9\,11} \\ 1\not{0}\not{1} \\ -\;\;44 \\ \hline 57 \end{array}$$

**Solve.**

**1.** 256 – (73 – 14) =
**256 – 59 = 197**

**2.** 256 – 73 – 14 =

**3.** 45 + (121 – 56) =

**4.** 45 + 121 – 56 =

**5.** 802 – (71 + 26) =

**6.** (802 – 71) + 26 =

**Use a calculator to solve.**

**7.** 24 + 195 – (84 – 32) =

**8.** 430 – 65 + (71 – 23) =

**9.** 56 + 604 – 122 + 118 =

**10.** 500 – (383 – 61) + 72 =

**Use two or more steps to answer the word problems.**

**11.** Vinnie began the day with $30. During the day, he spent $4 on lunch and $17 on books. How much money did Vinnie have left at the end of the day?

**12.** Alicia had a $20 bill, a $5 bill, and a $1 bill in her wallet. She used the $20 bill to buy gas that cost $7. How much money did Alicia have after she bought the gas?

**Check your answers on page 159.**

## Rounding and Estimating

You don't always need an exact answer to a problem. Sometimes you just need an answer that is close enough. In that case, you can **estimate** the answer. Estimating can also help you decide if your answer is reasonable.

One way to estimate answers is to round each number to its **lead digit**, or the digit that is the farthest to the left.

**Strategy** Estimate the sum: 489 + 35. Follow these steps.

**1.** Round each number to its lead digit.

489 ⟶ 500

35 ⟶ 40

**2.** Add the rounded numbers. 500 + 40 = 540

Estimate the difference: 1,864 – 628. Follow these steps.

**1.** Round each number to its lead digit.

1,864 ⟶ 2,000

628 ⟶ 600

**2.** Subtract the rounded numbers. 2,000 – 600 = 1,400

### Exercise 1: Round each number to its lead digit.

**1.** 376 ________

**2.** 88 ________

**3.** 2,213 ________

**4.** 824 ________

**5.** 33 ________

**6.** 655 ________

**7.** 3,766 ________

**8.** 573 ________

**Exercise 2: Round each number to the lead digit. Then add or subtract.**

**9.** $\begin{array}{r} 87 \\ -\ 62 \\ \hline \end{array}$

**10.** $\begin{array}{r} 48 \\ +\ 67 \\ \hline \end{array}$

**11.** $\begin{array}{r} 32 \\ +\ 85 \\ \hline \end{array}$

**12.** $\begin{array}{r} 323 \\ -\ \ 76 \\ \hline \end{array}$

**13.** $458 – 26 =

**14.** 841 + 67 =

**15.** 625 – 86 =

**16.** $\begin{array}{r} 817 \\ -\ 356 \\ \hline \end{array}$

**17.** $\begin{array}{r} 612 \\ +\ 178 \\ \hline \end{array}$

**18.** $\begin{array}{r} 415 \\ +\ 686 \\ \hline \end{array}$

**19.** $\begin{array}{r} \$8{,}235 \\ -\ \ \ \ 787 \\ \hline \end{array}$

**20.** 3,867 – 1,984 =

**21.** 5,235 + 488 =

**22.** 4,025 + 2,857 =

**Exercise 3: Answer the word problems. Round each number to its lead digit.**

**23.** A school computer lab bought a computer for $820 and a printer for $189. About how much money did the school spend?

**24.** Clark earned $382 last week and $423 this week. About how much did he earn in these two weeks?

**25.** The diameter of Earth is 7,927 miles. The diameter of the moon is 2,160 miles. About how much greater is the diameter of Earth than the diameter of the moon?

**26.** A car dealer has 78 cars and 23 trucks in stock. About how many more cars are in stock than trucks?

**Check your answers on page 159.**

## Selecting the Information

Sometimes, word problems contain some key facts along with other facts that are not important. When you solve a word problem, you have to select the information that you need.

Think about the question that is being asked. Look for the information that can help you answer that question. Ignore the information that will not help you answer the question.

**Strategy** Try this strategy on the example below. Use these steps.

**Step 1** Read the word problem.

**Step 2** Find the question. What is the question asking?

**Step 3** Circle or write down the facts that can help you answer the question.

**Step 4** Ignore or cross out the unnecessary information.

**Step 5** Solve the word problem.

**Example**

Andrea's car needed repairs. She spent $150 on brakes, $192 on tires, and $30 on an oil change. How much did Andrea spend on car repairs?

(1) $180
(2) $322
(3) $372
(4) $387

In Step 1 you read the problem. In Step 2 you found the question. It asks how much Andrea spent on repairs. In Step 3 you circled or wrote down the facts you need to answer the question. The facts are the amounts spent: $150, $192, and $30. In Step 4 you ignored or crossed out the extra information, such as the types of car repair. In Step 5 you solved the word problem by adding together the amounts. The correct answer is (3). Andrea spent $372 on car repairs.

## Practice

**Practice the strategy. Use the steps you learned. Solve each problem.**

**1.** Gina wrote a check for $140 to pay a plumber. The receipt below shows the balance in her account after the withdrawal. How much was in her account before she wrote the check?

| **Top Bank** | **Account # 0055806** |
|---|---|
| **Withdrawal Amount:** | **$140.00** |
| **Balance:** | **$165.00** |

(1) $165
(2) $305
(3) $25
(4) $205

**2.** David opened the new container of milk shown below. He poured 12 ounces of milk in a glass. How much milk is left in the container?

**1% Lowfat Milk**
**128 ounces**

(1) 136
(2) 140
(3) 100
(4) 116

**3.** A restaurant served 48 customers for breakfast, 67 customers for lunch, and 104 customers for dinner. How many customers were served in all that day?

(1) 115
(2) 171
(3) 219
(4) 152

**4.** Last month the Romeros spent $419 on groceries. This month the family spent $542 on groceries. How much more did the family spend on groceries this month than last month?

(1) $123
(2) $961
(3) $542
(4) $400

**5.** A store had 432 wireless telephones. After a 1-day sale, there were 278 phones left. How many phones were sold during the 1-day sale?

(1) 710
(2) 254
(3) 154
(4) 431

**6.** Cara drove 78 miles before she stopped for lunch. After lunch she drove 109 more miles. How many miles did Cara drive in all?

(1) 31
(2) 179
(3) 187
(4) 200

**Check your answers on page 159.**

# Unit 2 Wrap-up

Below are examples of the skills you learned in this unit. Read the examples and do the problems. Then check your answers.

**Examples.**

**1.** 4,368 + 521

```
   4,368      4,368      4,368      4,368
 +   521    +   521    +   521    +   521
       9         89        889      4,889
```

**2.** 364 + 2,797

```
       1         11        1 11       1 11
     364        364         364        364
 + 2,797    + 2,797     + 2,797    + 2,797
       1         61         161      3,161
```

**3.** 7,384 − 1,062

```
                                             Check:
   7,384      7,384      7,384      7,384      6,322
 − 1,062    − 1,062    − 1,062    − 1,062    + 1,062
       2         22        322      6,322      7,384
```

**4.** 543 − 274

```
                13        13       Check:
    3 13     4 3 13     4 3 1        1
     543        543       543       269
   − 274      − 274     − 274     + 274
       9         69       269       543
```

**5.** 334 − (55 + 14) − 26

```
            2 12 14     5 1
   55        334        265
 + 14      −  69      −  26
   69        265        239
```

**6.** Estimate: 943 + 89 + 1,549

```
    943   →     900
     89   →      90
 + 1,549  →  + 2,000
                2,990
```

**Problems.**

**1.** 6,012 + 23

**2.** 3,648 + 273

**3.** 1,306 − 702

**4.** 5,062 − 937

**5.** 825 − 247 − (73 + 62)

**6.** Estimate: 3,219 + 658 + 71

Check your answers on page 159.

# Unit 2 Practice

**Add.**

1. $\begin{array}{r} 64 \\ +\ 81 \\ \hline \end{array}$

2. $\begin{array}{r} 45 \\ +\ 37 \\ \hline \end{array}$

3. $\begin{array}{r} 567 \\ +\ 328 \\ \hline \end{array}$

4. $\begin{array}{r} 3,562 \\ +\ 784 \\ \hline \end{array}$

5. $\begin{array}{r} 370 \\ 4,219 \\ +\ \ 47 \\ \hline \end{array}$

6. $73 + 4 =$
7. $93 + 78 =$
8. $395 + 462 =$
9. $565 + 36 + 2,101 =$

**Subtract.**

10. $\begin{array}{r} 76 \\ -\ 24 \\ \hline \end{array}$

11. $\begin{array}{r} 81 \\ -\ 35 \\ \hline \end{array}$

12. $\begin{array}{r} 893 \\ -\ \ 46 \\ \hline \end{array}$

13. $\begin{array}{r} 703 \\ -\ 249 \\ \hline \end{array}$

14. $\begin{array}{r} 5,026 \\ -\ 1,273 \\ \hline \end{array}$

15. $58 - 5 =$
16. $70 - 16 =$
17. $415 - 162 =$
18. $7,580 - 237 =$

**Estimate. Round each number to the lead digit. Solve.**

19. $\begin{array}{r} 85 \\ +\ 194 \\ \hline \end{array}$

20. $\begin{array}{r} 841 \\ -\ 378 \\ \hline \end{array}$

21. $\begin{array}{r} 349 \\ +\ 6,512 \\ \hline \end{array}$

22. $\begin{array}{r} 279 \\ 525 \\ +\ 3,064 \\ \hline \end{array}$

23. $\begin{array}{r} 7,278 \\ -\ \ 853 \\ \hline \end{array}$

**Answer the word problems.**

24. Last year Harriet's company used 456 reams of white paper and 78 reams of yellow paper. How many more reams of white paper than yellow paper did her company use?

25. Tino works at a restaurant. One week he earned $160 in salary and $288 in tips. How much did Tino earn in all for that week?

**Check your answers on page 160.**

# Unit 2 GED Test Practice

**Read each problem carefully. Circle the number of the correct answer.**

1. Yoshi's car odometer reads 5,862 miles. She needs to drive 6,540 more miles. How many miles will Yoshi drive all together?
   (1) 12,402
   (2) 678
   (3) 10,678
   (4) 11,862

2. Bart has a ticket for the flight shown below. He has a voucher for $38 off the price. How much will Bart pay for the ticket?

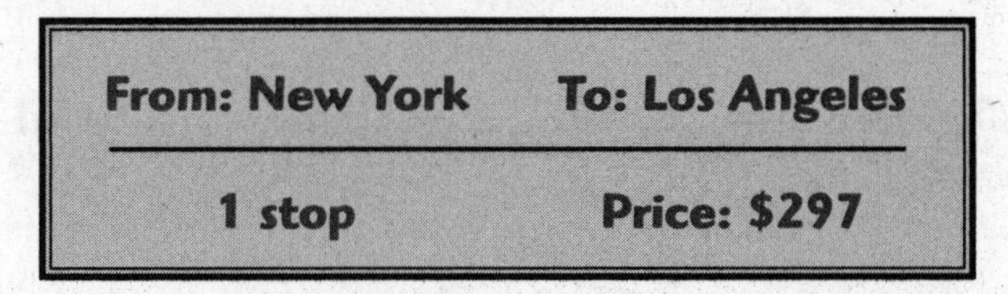


   (1) $159
   (2) $259
   (3) $269
   (4) $335

3. In a company, 136 people work in the warehouse, and 129 people work in the office. How many workers are there in all?
   (1) 265 people
   (2) 255 people
   (3) 245 people
   (4) 7 people

4. What operation is done first to solve 782 − (43 + 13) + 270?
   (1) 782 − 43
   (2) 43 + 13
   (3) 13 + 270
   (4) 782 + 270

5. A hardware store owner ordered 120 boxes of 1-inch screws, 180 boxes of 2-inch screws, and 40 boxes of 3-inch screws. How many boxes of screws were ordered in all?

(1) 650 boxes
(2) 450 boxes
(3) 340 boxes
(4) 310 boxes

6. A shoe store in the mall is 1,200 square feet. The storage area is 325 square feet, and the counter takes up 48 square feet. How much space is left for the shoe displays and seating?

(1) 1,573 square feet
(2) 1,152 square feet
(3) 875 square feet
(4) 827 square feet

7. Look at the bill shown below. What is the total amount due if it is paid by March 10?

**Hank's Home Supply Company**

**Total Amount Due: $420**

**Subtract $42 from the total if paid by March 10.**

(1) $378
(2) $380
(3) $388
(4) $462

8. Alex bought 48 square yards of carpet for the living room and 16 square yards for the bedroom. About how many square yards is this all together?

(1) 50 square yards
(2) 60 square yards
(3) 70 square yards
(4) 80 square yards

9. We took in 578 tons of glass and 243 tons of paper. What is the difference between the amount of paper and the amount of glass?

(1) 821 tons
(2) 345 tons
(3) 335 tons
(4) 325 tons

10. On Monday morning Tessa had 83 necklaces. She sold 15 necklaces during the day and made 8 more. How many necklaces did Tessa have at the end of the day?

(1) 60 necklaces
(2) 68 necklaces
(3) 75 necklaces
(4) 76 necklaces

**Check your answers on page 160.**

## Unit 2 Skill Check-Up Chart

Check your answers. In the first column, circle the numbers of any questions that you missed. Then look across the rows to see the skills you need to review and the pages where you can find each skill.

| Question | Skill | Page |
|---|---|---|
| 1 | Adding Larger Numbers | 29 |
| 2 | Subtracting with Renaming | 36 |
| 3 | Adding with Renaming | 30 |
| 4 | Order of Operations | 43 |
| 5 | Adding Three or More Numbers | 31 |
| 6, 10 | Addition and Subtraction: Two or More Steps | 42 |
| 7 | Renaming Zeros | 37 |
| 8 | Rounding and Estimating | 44 |
| 9 | Subtracting Larger Numbers | 35 |

# Unit 3 Multiplying and Dividing Whole Numbers

dividend

product

remainder

multiple

**In this unit you will learn about**

- multiplying and dividing numbers
- multiplying and dividing with renaming
- rounding and estimating answers
- using a calculator to multiply and divide
- solving word problems
- order of operations

You use multiplication and division when you are working with equal groups of numbers. You multiply whole numbers when you have many groups, each with the same amount. For example, you would multiply to find the number of eggs in three dozen.

Write an example of multiplying, such as finding the total price when buying five of the same item.

________________________________________

________________________________________

You divide whole numbers to find how many are in each group. For example, you and a friend might divide the price of lunch by two to see how much each of you should pay.

Write an example of dividing, such as equally dividing a batch of cookies among kids.

________________________________________

________________________________________

# Multiplication Facts

Use the 100 basic multiplication facts to multiply larger numbers. Complete this fact table, looking for patterns as you work. Check back to Unit 2 for help in using the columns and rows.

| × | 0 | 1 | 2 | 3 | 4 | 5 | 6 | 7 | 8 | 9 |
|---|---|---|---|---|---|---|---|---|---|---|
| 0 | 0 | 0 | 0 | 0 | 0 | 0 | | | | |
| 1 | 0 | 1 | 2 | 3 | 4 | | | | | |
| 2 | 0 | 2 | 4 | 6 | 8 | | | | | |
| 3 | 0 | 3 | 6 | 9 | | | | | | |
| 4 | 0 | 4 | 8 | | | | | | | |
| 5 | 0 | 5 | | | | | | | | |
| 6 | | | | | | | | | | |
| 7 | | | | | | | | | | |
| 8 | | | | | | | | | | |
| 9 | 0 | 9 | 18 | 27 | 36 | 45 | 54 | 63 | 72 | 81 |

**Complete the following multiplication facts.**

**1.** $\begin{array}{r} 5 \\ \underline{\times 4} \\ 20 \end{array}$ **2.** $\begin{array}{r} 6 \\ \underline{\times 7} \end{array}$ **3.** $\begin{array}{r} 3 \\ \underline{\times 0} \end{array}$ **4.** $\begin{array}{r} 2 \\ \underline{\times 6} \end{array}$ **5.** $\begin{array}{r} 7 \\ \underline{\times 5} \end{array}$ **6.** $\begin{array}{r} 4 \\ \underline{\times 3} \end{array}$

**7.** $\begin{array}{r} 7 \\ \underline{\times 4} \end{array}$ **8.** $\begin{array}{r} 3 \\ \underline{\times 7} \end{array}$ **9.** $\begin{array}{r} 4 \\ \underline{\times 0} \end{array}$ **10.** $\begin{array}{r} 5 \\ \underline{\times 9} \end{array}$ **11.** $\begin{array}{r} 8 \\ \underline{\times 2} \end{array}$ **12.** $\begin{array}{r} 1 \\ \underline{\times 1} \end{array}$

**13.** $\begin{array}{r} 6 \\ \underline{\times 8} \end{array}$ **14.** $\begin{array}{r} 3 \\ \underline{\times 3} \end{array}$ **15.** $\begin{array}{r} 7 \\ \underline{\times 2} \end{array}$ **16.** $\begin{array}{r} 8 \\ \underline{\times 8} \end{array}$ **17.** $\begin{array}{r} 3 \\ \underline{\times 2} \end{array}$ **18.** $\begin{array}{r} 8 \\ \underline{\times 9} \end{array}$

**19.** $8 \times 7 =$ **20.** $2 \times 1 =$ **21.** $9 \times 6 =$ **22.** $4 \times 8 =$ **23.** $3 \times 3 =$

**24.** $6 \times 8 =$ **25.** $5 \times 6 =$ **26.** $3 \times 7 =$ **27.** $7 \times 2 =$ **28.** $9 \times 4 =$

**Check your answers on page 161.**

# Multiplying by One-Digit Numbers

Use the multiplication facts to multiply larger numbers. Begin by multiplying the ones digits. Be sure to line up your answers in the correct column.

**Multiply:** 201 × 4

1. Line up the digits. Multiply the ones. 4 × 1 = 4 ones. Write 4 in the ones column.

$$\begin{array}{r} 201 \\ \times\ 4 \\ \hline 4 \end{array}$$

↑ **4 ones**

2. Multiply the tens. 4 × 0 = 0 tens. Write 0 in the tens column.

$$\begin{array}{r} 201 \\ \times\ 4 \\ \hline 04 \end{array}$$

↑ **0 tens**

3. Multiply the hundreds. 4 × 2 = 8 hundreds. Write 8 in the hundreds column.

$$\begin{array}{r} 201 \\ \times\ 4 \\ \hline 804 \end{array}$$

↑ **8 hundreds**

**Multiply.**

1. $\begin{array}{r} \$32 \\ \times\ 3 \\ \hline \$96 \end{array}$
2. $\begin{array}{r} \$10 \\ \times\ 2 \\ \hline \end{array}$
3. $\begin{array}{r} 21 \\ \times\ 4 \\ \hline \end{array}$
4. $\begin{array}{r} 13 \\ \times\ 1 \\ \hline \end{array}$
5. $\begin{array}{r} 153 \\ \times\ 1 \\ \hline \end{array}$
6. $\begin{array}{r} 101 \\ \times\ 6 \\ \hline \end{array}$
7. $\begin{array}{r} \$312 \\ \times\ 3 \\ \hline \end{array}$
8. $\begin{array}{r} 1{,}233 \\ \times\ 3 \\ \hline \end{array}$
9. $\begin{array}{r} 2{,}324 \\ \times\ 2 \\ \hline \end{array}$
10. $\begin{array}{r} 3{,}122 \\ \times\ 2 \\ \hline \end{array}$

11. \$43 × 2 = $\begin{array}{r} \$43 \\ \times\ 2 \\ \hline \$86 \end{array}$
12. 1,211 × 3 =
13. 34 × \$2 =
14. 121 × 4 =
15. 2,102 × 3 =
16. 12 × 4 =
17. 123 × 3 =
18. \$3,133 × 2 =

**Answer the word problems.**

19. There are 3 boxes on a shelf in a sporting goods store. Each box contains 12 baseballs. All together, how many baseballs are in the boxes?

20. A school bus has 32 seats. Each seat holds 2 children. How many children can be seated on the bus?

**Check your answers on page 161.**

# Multiplying by Two-Digit Numbers

When you multiply by a two-digit number, multiply each digit in the top number by each digit in the bottom number. You will get two partial products. A **partial product** is the total you get when you multiply a number by one digit of another number. Add the partial products to get the answer.

**Multiply:** 23 × 12

1. Line up the digits. Multiply by 2 ones. 2 × 3 = 6 ones. Write 6 in the ones column. 2 × 2 = 4 tens. Write 4 in the tens column.

```
 23
×12
 46  ← partial product
```

2. Multiply by 1 ten. 1 × 3 = 3 tens. Write 3 in the tens column under the 4. 1 × 2 = 2 hundreds. Write 2 in the hundreds column to the left of the 3.

```
 23
×12
 46
23← partial product
```

3. Add the partial products.

```
 23
×12
 46
+23
276
```

**Multiply.**

1. $32 × 21 = 32 + 64 = **$672**

```
  $32
 × 21
   32
 +64
 $672
```

2. $413 × 12

3. 13 × 13

4. 302 × 11

5. 133 × 12

6. $13 × 23

7. 1,202 × 14

8. $11 × 43

9. $12 × 11 =

```
  12
 ×11
  12
+12
$132
```

10. 304 × 12 =

11. $11 × 63 =

12. 210 × $14 =

**Answer the word problems.**

13. A group of 11 friends buys concert tickets. Tickets cost $24 each. Find the cost of tickets for the entire group.

14. If 120 customers shop at a supermarket each hour of the day, how many customers shop in a 24-hour day?

**Check your answers on page 161.**

# Multiplying by One-Digit Numbers with Renaming

When you multiply two digits, the product is sometimes 10 or more. When this happens, rename by carrying to the next column.

**Multiply:** $184 \times 3$

**1.** Line up the digits. Multiply the 4 by 3 ones. $4 \times 3 = 12$ ones. Rename 12 ones as 1 ten and 2 ones. Write 2 in the ones column. Carry 1 ten to the top of the next column.

$$\begin{array}{r} {\scriptstyle 1}\phantom{4} \\ 184 \\ \times\ \ 3 \\ \hline 2 \end{array}$$

**2.** Multiply the 8 tens by 3 ones. $8 \times 3 = 24$ tens. Then add the carried 1 ten. 24 tens + 1 ten = 25 tens. Rename 25 tens as 2 hundreds and 5 tens. Write 5 in the tens column. Carry 2 hundreds to the top of the next column.

$$\begin{array}{r} {\scriptstyle 2\,1}\phantom{4} \\ 184 \\ \times\ \ 3 \\ \hline 52 \end{array}$$

**3.** Multiply the 1 hundred by 3 ones. $1 \times 3 = 3$ hundreds. Then add the carried 2 hundreds. 3 hundreds + 2 hundreds = 5 hundreds. Write 5 in the hundreds column.

$$\begin{array}{r} {\scriptstyle 2\,1}\phantom{4} \\ 184 \\ \times\ \ 3 \\ \hline 552 \end{array}$$

**Multiply.**

**1.** $\begin{array}{r} {\scriptstyle 1\,1}\phantom{6} \\ 356 \\ \times\ \ 2 \\ \hline 712 \end{array}$

**2.** $\begin{array}{r} 14 \\ \times 7 \\ \hline \end{array}$

**3.** $\begin{array}{r} 2,140 \\ \times\ \ \ 3 \\ \hline \end{array}$

**4.** $\begin{array}{r} \$187 \\ \times\ \ 4 \\ \hline \end{array}$

**5.** $\begin{array}{r} 91 \\ \times 8 \\ \hline \end{array}$

**6.** $\begin{array}{r} \$1,372 \\ \times\ \ \ \ 5 \\ \hline \end{array}$

**7.** $\begin{array}{r} 905 \\ \times\ \ 9 \\ \hline \end{array}$

**8.** $\begin{array}{r} 83 \\ \times 2 \\ \hline \end{array}$

**9.** $\begin{array}{r} 526 \\ \times\ \ 7 \\ \hline \end{array}$

**10.** $\begin{array}{r} 1,315 \\ \times\ \ \ 3 \\ \hline \end{array}$

**11.** $15 \times 4 =$

$\begin{array}{r} {\scriptstyle 2}\phantom{5} \\ 15 \\ \times 4 \\ \hline 60 \end{array}$

**12.** $6,012 \times 2 =$

**13.** $828 \times \$5 =$

**14.** $2,130 \times 7 =$

**Answer the word problems.**

**15.** Each day, the cafeteria needs 130 cartons of milk. The cafeteria is open 5 days a week. How many cartons of milk does it need each week?

**16.** There are 6 people in an elevator. Each person weighs 145 pounds. All together, the people in the elevator weigh how much?

**Check your answers on page 162.**

# Multiplying by Two-Digit Numbers with Renaming

You may need to rename several times when multiplying by 2-digit multipliers.

**Multiply:** 1,406 × 53

**1.** Line up the digits.

```
  1,406
×    53
```

**2.** Multiply by 3 ones. Multiply by 5 tens.

```
  2  3
  1  1
 1,406
×   53
  4218
  7030
```

**3.** Add the partial products.

```
  1,406
×    53
   4218
 +7030
 74,518
```

**Multiply.**

**1.**
```
   1
   2
  47
 ×23
 141
 +94
1,081
```

**2.**
```
  1,806
×    37
```

**3.**
```
  $421
×   56
```

**4.**
```
  2,307
×    48
```

**5.**
```
  $718
×   45
```

**6.**
```
  $24
×  61
```

**7.**
```
  3,219
×    72
```

**8.**
```
  $95
×  38
```

**9.** 903 × $14 =
```
     1
    903
 × $14
   3612
 + 903
$12,642
```

**10.** 8,215 × 51 =

**11.** 70 × 32 =

**Answer the word problems.**

**12.** The auditorium seats 1,450 people. A play scheduled for 15 performances has completely sold out. How many tickets have been sold?

**13.** The Greenway Recycling Club estimates that it collects 2,500 cans each month. At this rate, how many cans does the club collect each year? (Hint: There are 12 months in a year.)

**Check your answers on page 162.**

# Multiplying by Multiples of 10

You can multiply by 10, 100, or multiples of 10 without writing a row of partial products with zeros.

**Multiply:** 3,425 × 100

**1.** Multiply by 0 ones. Write zero in the ones column.

```
  3,425
×   100
      0
```

**2.** Multiply by 0 tens. Write zero in the tens column.

```
  3,425
×   100
     00
```

**3.** Multiply by 1 hundred. Write the answer to the left of the zeros. The answer is the top number plus two zeros.

```
  3,425
×   100
342,500
```

**Multiply.**

**1.**
```
   61
×  40
2,440
```

**2.**
```
  4,231
×   100
```

**3.**
```
  $213
×  300
```

**4.**
```
  $735
×  100
```

**5.**
```
  1,812
×    70
```

**6.** 311 × 50 = 15,550

```
   311
×   50
15,550
```

**7.** 161 × $700 =

**8.** 103 × 90 =

**9.** $25 × 20 =

**10.** 67 × 30 =

**11.** 420 × 600 =

**12.** 3,306 × 50 =

**13.** 833 × 200 =

**Answer the word problems.**

**14.** There are 150 reams of paper in an office supply room. Each ream has 500 sheets of paper. How many sheets of paper are in the supply room?

**15.** There are 80 boxes stacked on a pallet. Each box contains 12 bottles of shampoo. How many bottles of shampoo are on the pallet?

**Check your answers on page 162.**

# Division Facts

You can use a multiplication table to complete division facts. Look at the division fact to the right.

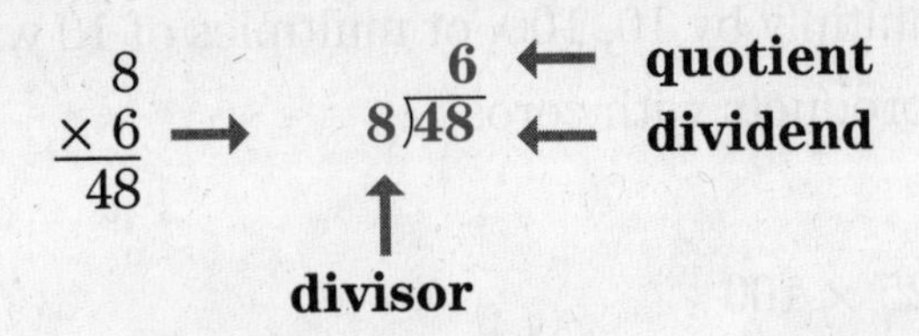


To find the answer to the division fact, first find the row that begins with the divisor, 8. Next, move to the right until you find the dividend, 48. Then, move up the column to the top of the chart to find the answer, 6.

| ÷ | 0 | 1 | 2 | 3 | 4 | 5 | 6 | 7 | 8 | 9 |
|---|---|---|---|---|---|---|---|---|---|---|
| 1 | 0 | 1 | 2 | 3 | 4 | 5 | 6 | 7 | 8 | 9 |
| 2 | 0 | 2 | 4 | 6 | 8 | 10 | 12 | 14 | 16 | 18 |
| 3 | 0 | 3 | 6 | 9 | 12 | 15 | 18 | 21 | 24 | 27 |
| 4 | 0 | 4 | 8 | 12 | 16 | 20 | 24 | 28 | 32 | 36 |
| 5 | 0 | 5 | 10 | 15 | 20 | 25 | 30 | 35 | 40 | 45 |
| 6 | 0 | 6 | 12 | 18 | 24 | 30 | 36 | 42 | 48 | 54 |
| 7 | 0 | 7 | 14 | 21 | 28 | 35 | 42 | 49 | 56 | 63 |
| 8 | 0 | 8 | 16 | 24 | 32 | 40 | 48 | 56 | 64 | 72 |
| 9 | 0 | 9 | 18 | 27 | 36 | 45 | 54 | 63 | 72 | 81 |

**Complete the following division facts.**

**1.** $5\overline{)30}$ (answer: **6**)  **2.** $9\overline{)72}$  **3.** $2\overline{)6}$  **4.** $7\overline{)35}$  **5.** $3\overline{)12}$

**6.** $8\overline{)16}$  **7.** $1\overline{)2}$  **8.** $1\overline{)8}$  **9.** $4\overline{)36}$  **10.** $5\overline{)35}$

**11.** $4\overline{)20}$  **12.** $8\overline{)48}$  **13.** $7\overline{)21}$  **14.** $6\overline{)6}$  **15.** $7\overline{)56}$

**16.** $42 \div 6 =$ **7**  **17.** $45 \div 5 =$  **18.** $1 \div 1 =$  **19.** $10 \div 2 =$

**20.** $24 \div 8 =$  **21.** $20 \div 4 =$  **22.** $15 \div 5 =$  **23.** $54 \div 9 =$

**24.** $28 \div 7 =$  **25.** $20 \div 5 =$  **26.** $32 \div 4 =$  **27.** $12 \div 4 =$

**Check your answers on page 162.**

# Dividing by One-Digit Numbers

Use the division facts when you divide. Be sure to put each answer in the correct column.

**Divide:** $186 \div 3$

**1.** Divide. Since you can't divide 1 by 3, divide 18 by 3. $18 \div 3 = 6$. Write the 6 above the 8.

$$3\overline{)186}^{\ \ 6}$$

**2.** Divide again. $6 \div 3 = 2$. Write the 2 above the 6.

$$3\overline{)186}^{\ 62}$$

**3.** Check by multiplying the answer (62) by the divisor (3).

$$3\overline{)186}^{\ 62} \longleftrightarrow \begin{array}{r} 62 \\ \times\ 3 \\ \hline 186 \end{array}$$

**Divide. Use multiplication to check your answers.**

1. $2\overline{)148}^{\ 74}$ $\quad \begin{array}{r} 74 \\ \times\ 2 \\ \hline 148 \end{array}$
2. $3\overline{)\$273}$
3. $8\overline{)88}$
4. $5\overline{)2{,}055}$
5. $5\overline{)555}$
6. $6\overline{)3{,}066}$
7. $6\overline{)486}$
8. $7\overline{)\$147}$
9. $156 \div 3 =$
10. $\$284 \div 4 =$
11. $1{,}684 \div 4 =$
12. $186 \div 2 =$

**Answer the word problems. Use multiplication to check your answers.**

13. A charity group has 88 presents to give to 8 children. Each child will receive the same number of presents. How many presents will each child receive?

14. An apple orchard gives 2 free apples to each customer. Last weekend 128 apples were given away. How many customers went to the apple orchard last weekend?

**Check your answers on page 163.**

# Remainders

Sometimes you will have an amount left over after you have finished dividing two numbers. The amount left over is called the **remainder**. A remainder is part of the answer. Use the letter R to represent a remainder.

$$\begin{array}{r} 2 \text{ R1} \\ 4\overline{)9} \\ -8 \\ \hline 1 \end{array}$$

**Divide:** 189 ÷ 4

**1.** Divide. Since 18 ÷ 4 is not a basic fact, use the next lower basic fact, 16 ÷ 4, instead. 16 ÷ 4 = 4, so 18 ÷ 4 = 4 with an amount left over. Write the 4 above the 8. Multiply. 4 × 4 = 16. Write the 16 under the 18 and subtract. The amount left is 2.

$$\begin{array}{r} 4\phantom{9} \\ 4\overline{)189} \\ -16\phantom{9} \\ \hline 2\phantom{9} \end{array}$$

**2.** Divide again by bringing down the next digit, 9. The 9 brought down beside the 2 makes 29. 29 ÷ 4 = 7 with an amount left over. Multiply. 7 × 4 = 28. Write the 28 under the 29 and subtract. 29 − 28 = 1. The remainder is 1.

$$\begin{array}{r} 47 \text{ R1} \\ 4\overline{)189} \\ -16\downarrow \\ \hline 29 \\ -28 \\ \hline 1 \end{array}$$

**3.** Check your answer by multiplying. Add the remainder.

$$\begin{array}{r} \overset{2}{4}7 \\ \times\ 4 \\ \hline 188 \\ +\ \ 1 \\ \hline 189 \end{array}$$

**Divide. Use multiplication to check your answers.**

**1.** $$\begin{array}{r} 27 \text{ R5} \\ 6\overline{)167} \\ -12\phantom{7} \\ \hline 47 \\ -42 \\ \hline 5 \end{array} \qquad \begin{array}{r} \overset{4}{2}7 \\ \times\ 6 \\ \hline 162 \\ +\ \ 5 \\ \hline 167 \end{array}$$

**2.** $4\overline{)295}$

**3.** $7\overline{)317}$

**4.** $2\overline{)35}$

**5.** $3\overline{)71}$

**6.** $5\overline{)493}$

**7.** $9\overline{)257}$

**8.** $8\overline{)99}$

**9.** 187 ÷ 4 =

**10.** 87 ÷ 7 =

**11.** 98 ÷ 6 =

**12.** 228 ÷ 9 =

Check your answers on page 163.

# Zeros in Division

In some problems, the dividend may have one or more zeros. Bring down the next digit, even if it is a zero. When dividing into zero, write zero in the answer.

**Divide:** 1,302 ÷ 3

**1.** Divide. 13 ÷ 3 = 4 with an amount left over. Multiply, subtract, and bring down the 0.

$$\begin{array}{r} 4\phantom{02} \\ 3\overline{)1{,}302} \\ -1\,2\downarrow\phantom{2} \\ \hline 10\phantom{2} \end{array}$$

**2.** Divide. 10 ÷ 3 = 3 with an amount left over. Multiply and subtract. Bring down the 2.

$$\begin{array}{r} 43\phantom{2} \\ 3\overline{)1{,}302} \\ -1\,2\phantom{02} \\ \hline 10\phantom{2} \\ -\ 9\downarrow \\ \hline 12 \end{array}$$

**3.** Divide. 12 ÷ 3 = 4. Multiply and subtract. There is no remainder. Check your answer.

$$\begin{array}{r} 434 \\ 3\overline{)1{,}302} \\ -1\,2\phantom{02} \\ \hline 10\phantom{2} \\ -\ 9\phantom{2} \\ \hline 12 \\ -12 \\ \hline 0 \end{array} \qquad \begin{array}{r} {}^{1\,1}\phantom{4} \\ 434 \\ \times\ \ 3 \\ \hline 1{,}302 \end{array}$$

**Divide. Use multiplication to check your answers.**

**1.**
$$\begin{array}{r} 402\text{ R1} \\ 3\overline{)1207} \\ -12\phantom{07} \\ \hline 00\phantom{7} \\ -\ 0\phantom{7} \\ \hline 07 \\ -\ 6 \\ \hline 1 \end{array} \qquad \begin{array}{r} 402 \\ \times\ \ 3 \\ \hline 1{,}206 \\ +\ \ 1 \\ \hline 1{,}207 \end{array}$$

**2.** $4\overline{)2{,}432}$

**3.** $6\overline{)3{,}083}$

**4.** $7\overline{)1{,}449}$

**5.** 1,808 ÷ 2 =

**6.** \$2,550 ÷ 3 =

**7.** 1,062 ÷ 9 =

**8.** 1,608 ÷ 8 =

**Answer the word problems. Use multiplication to check your answers.**

**9.** A supermarket received 600 tomatoes. The tomatoes are in groups of 4. How many groups of tomatoes are there?

**10.** A total of 2,400 guests were served at 3 dinners. Each dinner served the same number of guests. How many guests were served at each dinner?

Check your answers on page 163.

# Dividing by Two-Digit Numbers

Estimation can help you divide by two-digit numbers.

**Divide:** 807 ÷ 25

**1.** Estimate how many times the first digit of 25 (2) goes into the first digit of 807 (8). Because 2 goes into 8 exactly 4 times, try 4. 4 × 25 = 100. 100 is greater than 80, so 4 is too large.

$$\begin{array}{r} 4\phantom{07} \\ 25\overline{)807} \\ -100\phantom{0} \end{array}$$

**2.** Try the next smaller number, 3. 3 × 25 = 75. Subtract 75 from 80. 80 − 75 = 5. Bring down the 7.

$$\begin{array}{r} 3\phantom{7} \\ 25\overline{)807} \\ -75\downarrow \\ \hline 57 \end{array}$$

**3.** Decide how many times the first digit of 25 (2) goes into the first digit of 57 (5). Because 2 goes into 5 about 2 times, try 2. 2 × 25 = 50. Since 50 is less than 57, subtract. 57 − 50 = 7. The remainder is 7. Check your answer.

$$\begin{array}{r} 32 \text{ R7} \\ 25\overline{)807} \\ -75\phantom{7} \\ \hline 57 \\ -50 \\ \hline 7 \end{array} \qquad \begin{array}{r} \overset{1}{32} \\ \times 25 \\ \hline \overset{1}{160} \\ +64\phantom{0} \\ \hline 800 \\ +\phantom{0}7 \\ \hline 807 \end{array}$$

**Divide. Use multiplication to check your answers.**

1. $\begin{array}{r} 14 \text{ R2} \\ 39\overline{)548} \\ -39\phantom{8} \\ \hline 158 \\ -156 \\ \hline 2 \end{array} \qquad \begin{array}{r} \overset{1}{\overset{3}{14}} \\ \times 39 \\ \hline 126 \\ +\phantom{0}42\phantom{0} \\ \hline 546 \\ +\phantom{00}2 \\ \hline 548 \end{array}$

2. $24\overline{)771}$

3. $53\overline{)639}$

4. $12\overline{)558}$

5. 912 ÷ 43 =

6. 864 ÷ 61 =

7. $962 ÷ 26 =

8. 930 ÷ 34 =

**Answer the word problems. Use multiplication to check your answers.**

9. 14 employees worked a total of 490 hours. Each employee worked the same number of hours. How many hours did each employee work?

10. A textbook contains 330 pages. The book has chapters that are each 22 pages long. How many chapters are in the textbook?

**Check your answers on page 164.**

# Order of Operations

Some problems ask you to use more than one operation. Always follow this order of operations to solve the problem.

**Step 1:** Do any operations in parentheses ( ) first.
**Step 2:** Multiply and divide, working from left to right.
**Step 3:** Add and subtract, working from left to right.

**Solve:** $8 + 6 \div 2$

**1.** First, do any operations shown in ( ). There are none. Move to the next step.

$\mathbf{8 + 6 \div 2 =}$

**2.** Do all multiplication and division. There is one division step. $6 \div 2 = 3$.

$\mathbf{8 + 6 \div 2 =}$
$\mathbf{8 + \quad 3 \quad =}$

**3.** Do all addition and subtraction. There is one addition step. $8 + 3 = 11$.

$\mathbf{8 + 3 =}$
$\mathbf{8 + 3 = 11}$

**Solve.**

**1.** $(4 + 5) \times 2 =$
$\mathbf{9 \quad \times 2 = 18}$

**2.** $4 + 5 \times 2 =$
$\mathbf{4 + \quad 10 \quad = 14}$

**3.** $6 + 2 \times 1 =$

**4.** $7 - 4 + 2 =$

**5.** $10 \div (2 + 3) =$

**6.** $5 \times 4 - 1 =$

**7.** $8 + 2 \div 2 =$

**8.** $18 \div 6 - 3 =$

**9.** $(14 - 8) \times 1 =$

**10.** $12 + 6 \div 2 =$

**11.** $21 + 9 \div 3 =$

**12.** $15 \times 2 - 1 =$

**13.** $12 \div 6 - 2 =$

**14.** $16 \div (4 + 4) =$

**15.** $(7 + 0) \times 7 =$

**16.** $9 \times 5 - 5 =$

**17.** $(24 - 12) \times 2 =$

**18.** $30 + 6 \div 2 =$

**19.** $14 + 8 \div 2 =$

**20.** $33 + 12 \div 3 =$

**Check your answers on page 164.**

## Rounding and Estimating

Sometimes you don't need an exact answer to a multiplication or division problem. You can estimate to find an answer that's close enough. First round the numbers, and then multiply or divide the rounded numbers.

**Strategy** Estimate the product: 92 × 28.

1. Round each number to its lead digit, or the first digit on the left. 92 → 90 28 → 30
2. Multiply the rounded numbers. 90 × 30 = 2,700

Estimate the quotient: 2,349 ÷ 53.

1. Round each number. Use basic division facts to help. Think 20 ÷ 5. 2,349 → 2,000 53 → 50
2. Divide the rounded numbers. 2,000 ÷ 50 = 40

**Exercise 1: Round the numbers to make easy pairs to work with. Use basic facts to help. Do not solve.**

1. 415 ÷ 7 ______________________
2. 69 × 8 ______________________
3. 3,110 ÷ 41 ______________________
4. 808 × 88 ______________________
5. 5,566 ÷ 72 ______________________
6. 6,245 × 33 ______________________

**Exercise 2: Round each number to the lead digit. Rewrite the problem. Then multiply.**

7. $\begin{array}{r} 93 \\ \times 41 \\ \hline \end{array} \rightarrow \begin{array}{r} 93 \\ \times \quad 40 \\ \hline 3{,}600 \end{array}$

8. $\begin{array}{r} 86 \\ \times 27 \\ \hline \end{array}$

9. $\begin{array}{r} 72 \\ \times 49 \\ \hline \end{array}$

10. $\begin{array}{r} 68 \\ \times 32 \\ \hline \end{array}$

11. $284 \times 93 =$

12. $764 \times 93 =$

13. $459 \times 36 =$

14. $917 \times 35 =$

**Exercise 3: Round each number. Rewrite the problem. Then divide. Use basic facts to help.**

15. $54\overline{)362}$

$\begin{array}{r} 8 \\ 50\overline{)400} \\ -400 \\ \hline 0 \end{array}$

16. $77\overline{)417}$

17. $42\overline{)245}$

18. $53\overline{)1{,}579}$

19. $377 \div 75 =$

20. $964 \div 52 =$

21. $3{,}120 \div 58 =$

22. $2{,}642 \div 33 =$

**Exercise 4: Answer the word problems.**

23. Fast-Oil changed the oil for 39 cars in 1 day. At this rate, estimate the number of cars Fast-Oil will service in 312 days.

24. Rex bought a used car for $2,349. He will pay for the car in 12 equal payments. Estimate the amount of each payment.

Check your answers on page 164.

## Using a Calculator to Multiply and Divide

A calculator can help you multiply and divide. Always clear your calculator with the AC key before you begin. Check each number after you enter it. You can also estimate to check your answer.

**Strategy** Learn how to use the Casio *fx-260* calculator to multiply. Follow these steps.

**Multiply 923 × 42.**

| | Press the key. | Read the display. |
|---|---|---|
| **1.** Clear the calculator. | → AC | → 0. |
| **2.** Enter the first number. | → 9 2 3 | → 923. |
| **3.** Press the multiplication key. | → × | → 923. |
| **4.** Enter the next number. | → 4 2 | → 42. |
| **5.** Press the equals key. | → = | → 38766. |

**6.** Estimate to check. 900 × 40 = 36,000

**Strategy** Learn how to use the Casio *fx-260* calculator to divide. Follow these steps.

**Divide 378 ÷ 18.**

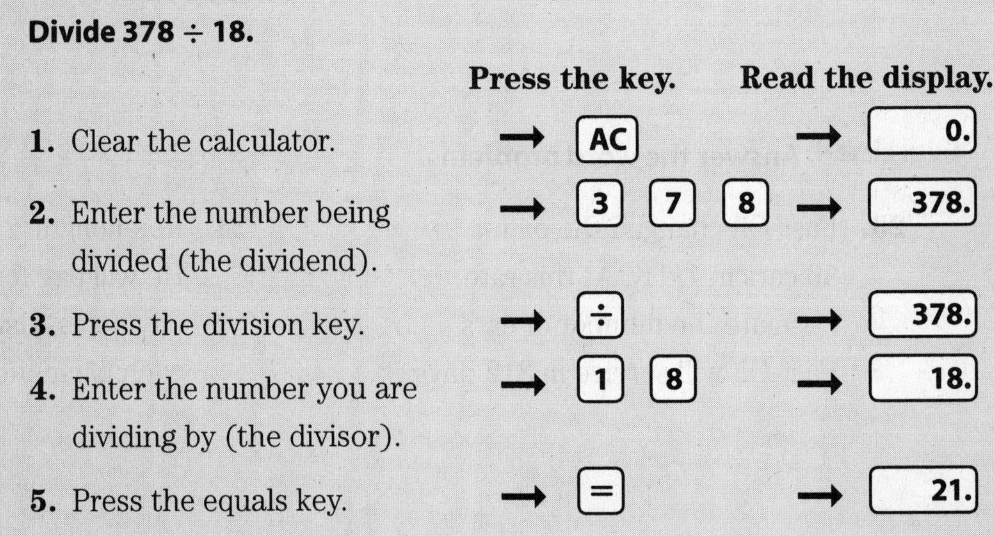

| | Press the key. | Read the display. |
|---|---|---|
| **1.** Clear the calculator. | → AC | → 0. |
| **2.** Enter the number being divided (the dividend). | → 3 7 8 | → 378. |
| **3.** Press the division key. | → ÷ | → 378. |
| **4.** Enter the number you are dividing by (the divisor). | → 1 8 | → 18. |
| **5.** Press the equals key. | → = | → 21. |

**6.** Estimate to check. 400 × 20 = 20

**Exercise 1: Use a calculator to multiply. Estimate to check.**

1. $\begin{array}{r} 46 \\ \times 35 \\ \hline \end{array}$

2. $\begin{array}{r} 59 \\ \times 24 \\ \hline \end{array}$

3. $\begin{array}{r} 62 \\ \times 49 \\ \hline \end{array}$

4. $\begin{array}{r} 86 \\ \times 75 \\ \hline \end{array}$

5. $318 \times 54 =$

6. $1{,}263 \times 219 =$

7. $581 \times 188 =$

8. $2{,}046 \times 362 =$

**Exercise 2: Use a calculator to divide. Estimate to check.**

9. $28\overline{)896}$

10. $32\overline{)\$576}$

11. $19\overline{)399}$

12. $53\overline{)2{,}650}$

13. $1{,}458 \div 6 =$

14. $3{,}225 \div 5 =$

15. $1{,}840 \div 40 =$

16. $\$1{,}092 \div 13 =$

**Exercise 3: Use a calculator to solve. Estimate to check.**

17. A wedding gift cost $78. Each of three friends will share the cost equally. How much will each person pay?

18. A jar of peanut butter weighs 36 ounces. How many ounces does a case of 24 peanut butter jars weigh?

19. Jorge earns $1,460 a month. How much does he earn in 12 months?

**Check your answers on page 165.**

## Choosing the Operation

On the GED Mathematics Test, you solve word problems. When you solve a word problem, you have to decide which **operation** to use—addition, subtraction, multiplication, or division.

One strategy for passing the GED Test is to look for key words. Some word problems have key words that can help you decide which operation to use. Here are some key word examples:

| Addition | Subtraction | Multiplication | Division |
|---|---|---|---|
| all together<br>and<br>in all<br>sum<br>total | difference<br>less than<br>more than<br>remain<br>decrease | in all<br>product<br>total<br>as many, as much<br>twice | average<br>each<br>equally<br>every<br>half |

**Strategy** Try this strategy on the example below. Use these steps.

**Step 1** Read the word problem.

**Step 2** Look for key words.

**Step 3** Use the key words to help decide which operation to use.

**Step 4** Solve the word problem.

**Example**

A small box of cereal weighs 24 ounces. A large box weighs 72 ounces. How much do the boxes weigh in all?

(1) 96

(2) 72

(3) 1,152

(4) 48

In Step 1 you read the problem. In Step 2 you found the key words. The key words are *in all.* In Step 3 you used the key words to help decide which operation to use. You should use addition. In Step 4 you solved the problem. The correct answer is (1). The boxes weigh 96 ounces in all.

## Practice

**Practice the strategy. Use the steps you learned. Solve the problems.**

**1.** Angela has 3 coupons that are each worth $5 off. What is the total amount she will save if she uses all 3 coupons?

· COUPON ·

**$5.00 OFF the regular price**

Expires 9/30

(1) $5
(2) $8
(3) $10
(4) $15

**2.** Eric bought a carton of 12 eggs. After arriving home, he found that 3 eggs were broken. How many eggs were not broken?

(1) 15
(2) 4
(3) 9
(4) 3

**3.** On Monday, Roz spent $35 at the grocery store. She spent about 3 times as much on Friday. How much did Roz spend on Friday?

(1) $38
(2) $105
(3) $32
(4) $70

**4.** A slice of apple pie contains 350 calories. If 2 people each eat half a slice, how many calories has each person eaten?

(1) 175
(2) 348
(3) 350
(4) 700

**5.** A pound of ham is $4, and a pound of cheese is $3. How much will it cost all together to buy a pound of each?

(1) $1
(2) $7
(3) $10
(4) $12

**6.** Tom bought 15 items at the grocery store. Marita bought 45 items. How many more items did Marita buy than Tom?

(1) 675
(2) 3
(3) 60
(4) 30

**Check your answers on page 165.**

# Unit 3 Wrap-up

Below are examples of the skills in this unit. Read the examples and work the problems. Then check your answers.

**Examples.**

**1.** 231 × 3

$$\begin{array}{r} 231 \\ \times\ \ 3 \\ \hline 3 \end{array} \qquad \begin{array}{r} 231 \\ \times\ \ 3 \\ \hline 93 \end{array} \qquad \begin{array}{r} 231 \\ \times\ \ 3 \\ \hline 693 \end{array}$$

**2.** 193 × 4

$$\begin{array}{r} {\scriptstyle 1}\ \ \\ 193 \\ \times\ \ 4 \\ \hline 2 \end{array} \qquad \begin{array}{r} {\scriptstyle 3\,1}\ \ \\ 193 \\ \times\ \ 4 \\ \hline 72 \end{array} \qquad \begin{array}{r} {\scriptstyle 3\,1}\ \ \\ 193 \\ \times\ \ 4 \\ \hline 772 \end{array}$$

**3.** 268 × 100

$$\begin{array}{r} 268 \\ \times\ 100 \\ \hline 0 \end{array} \qquad \begin{array}{r} 268 \\ \times\ 100 \\ \hline 00 \end{array} \qquad \begin{array}{r} 268 \\ \times\ 100 \\ \hline 26{,}800 \end{array}$$

**4.** Round and estimate: 193 × 42. Then find the exact answer.

$$\begin{array}{r} 193 \\ \times\ 42 \\ \hline \end{array} \begin{array}{c} \rightarrow \\ \rightarrow \end{array} \qquad \begin{array}{r} 200 \\ \times\ 40 \\ \hline 8{,}000 \end{array} \qquad \begin{array}{r} {\scriptstyle 3\,1} \\ {\scriptstyle 1}\ \ \\ 193 \\ \times\ 42 \\ \hline 386 \\ +\ 772\ \ \\ \hline 8{,}106 \end{array}$$

**5.** 108 ÷ 5 Use multiplication to check your answer.

$$\begin{array}{r} 2\ \ \\ 5\overline{)108} \\ -10\ \ \\ \hline 0\ \ \end{array} \qquad \begin{array}{r} 21\ \text{R3} \\ 5\overline{)108}\ \ \ \ \ \\ 10\ \ \ \ \ \ \\ \hline 08\ \ \ \ \ \\ -\ 05\ \ \ \ \ \\ \hline 3\ \ \ \ \ \end{array} \qquad \begin{array}{r} 21 \\ \times\ 5 \\ \hline 105 \\ +\ \ 3 \\ \hline 108 \end{array}$$

**6.** 7 + 3 × 2

**7 + 6 = 13**

**Problems.**

**1.** 201 × 4

**2.** 157 × 4

**3.** 516 × 100

**4.** Round and estimate: 323 × 57. Then find the exact answer.

$$\begin{array}{r} 323 \\ \times\ 57 \\ \hline \end{array} \begin{array}{c} \rightarrow \\ \rightarrow \end{array}$$

**5.** 217 ÷ 8 Use multiplication to check your answer.

$8\overline{)217}$

**6.** 10 ÷ 2 − 1 =

**Check your answers on page 165.**

## Unit 3 Practice

**Multiply or divide.**

**1.** $6 \times 7 =$  **2.** $7 \times 5 =$  **3.** $9 \times 3 =$  **4.** $9 \times 2 =$

**5.** $16 \div 4 =$  **6.** $24 \div 8 =$  **7.** $32 \div 4 =$  **8.** $63 \div 7 =$

**Multiply.**

**9.** $\begin{array}{r} 48 \\ \underline{\times\ 6} \end{array}$  **10.** $\begin{array}{r} 512 \\ \underline{\times\ \ 4} \end{array}$  **11.** $\begin{array}{r} 26 \\ \underline{\times 32} \end{array}$  **12.** $\begin{array}{r} 403 \\ \underline{\times\ 81} \end{array}$

**13.** $52 \times 8 =$  **14.** $5 \times 911 =$  **15.** $40 \times 84 =$  **16.** $3{,}452 \times 60 =$

**Divide.**

**17.** $5\overline{)135}$  **18.** $4\overline{)93}$  **19.** $6\overline{)4{,}027}$  **20.** $14\overline{)589}$

**21.** $96 \div 6 =$  **22.** $308 \div 7 =$  **23.** $1{,}839 \div 5 =$  **24.** $7{,}843 \div 32 =$

**Solve.**

**25.** $(75 - 30) \div 9 =$  **26.** $11 + 6 \times 8 =$  **27.** $65 - 10 \div 2 =$  **28.** $6 + 4 \times 3 =$

**Check your answers on page 165.**

# Unit 3 GED Test Practice

**Read each problem carefully. Circle the number of the correct answer.**

**1.** A carton includes 42 packages of pens. There are 3 pens in each package. How many pens are there in all?

(1) 14
(2) 39
(3) 45
(4) 126

**2.** A job fair splits 306 people equally into 3 different workshops. How many people will be in each workshop?

(1) 309
(2) 303
(3) 102
(4) 12

**3.** A company paid $817 in total overtime pay to 19 different employees. Each employee earned the same amount of overtime pay. How much did each employee earn?

(1) $43
(2) $798
(3) $817
(4) $15,223

**4.** An office manager orders 144 boxes of the paper clips shown below. How many paper clips are there in all?

(1) 56
(2) 344
(3) 14,400
(4) 28,800

**5.** A store manager ordered a total of 280 golf balls. How many boxes did he order?

(1) 2,240
(2) 288
(3) 272
(4) 35

**6.** A sporting goods store makes $12 on each hockey stick it sells. How much does the store make if it sells 110 sticks?

(1) $1,322
(2) $1,320
(3) $330
(4) $122

**7.** Ty pays $296 each month for his car payment. If the loan is for 48 months, how much will he pay for the car all together?

(1) $14,208
(2) $3,552
(3) $1,184
(4) $344

**8.** A coach has 38 mouth guards to give out to her 16 players. If each player gets an equal amount, how many mouth guards will be left over?

(1) 22
(2) 6
(3) 5
(4) 2

9. What is the difference in this equation?

$(2 \times 19) - (3 \times 12)$

(1) 2
(2) 34
(3) 384
(4) 420

10. During a sale, a store sold 108 books at $19 each. What is the best estimate of the total amount of book sales?

(1) $2,000
(2) $1,500
(3) $1,000
(4) $900

**Check your answers on page 166.**

## Unit 3 Skill Check-Up Chart

Check your answers. In the first column, circle the numbers of any questions that you missed. Then look across the rows to see the skills you need to review and the pages where you can find each skill.

| Question | Skill | Page |
|---|---|---|
| 1 | Multiplying by One-Digit Numbers | 55 |
| 2 | Zeros in Division | 63 |
| 3 | Dividing by Two-Digit Numbers | 64 |
| 4 | Multiplying by Multiples of 10 | 59 |
| 5 | Dividing by One-Digit Numbers | 61 |
| 6 | Multiplying by Two-Digit Numbers | 56 |
| 7 | Multiplying by Two-Digit Numbers with Renaming | 58 |
| 8 | Remainders | 62 |
| 9 | Order of Operations | 65 |
| 10 | Rounding and Estimating | 66–67 |

# Unit 4 Decimals and Percents

decimal

part

zero

percent

**In this unit you will learn about**

- identifying place values in decimals
- comparing, rounding, and estimating decimals
- adding and subtracting decimals
- multiplying and dividing decimals
- writing percents as decimals
- finding the part and percent
- solving word problems with decimals
- solving word problems with percents

Decimals are numbers used to show amounts less than 1 or amounts between two whole numbers.

You often see decimals being used. You see decimals in the weights on packages of meat and in the price of gas. Write two examples of decimals you are familiar with, such as the price of a gallon of milk.

________________________________________

________________________________________

Another way to show part of a whole is with a percent. In a percent, a number is compared to 100 (50 out of 100 is 50%).

Write an example of a percent, such as the percent off an item on sale.

________________________________________

________________________________________

# Decimal Place Names and Values

Like whole numbers, **decimals** are made up of digits with different place values. A decimal shows part of a whole number. Numbers to the right of the decimal point are used to show amounts less than 1.

In the decimal 45.017, the digit 4 is in the tens place, the digit 5 is in the ones place, the digit 0 is in the tenths place, the digit 1 is in the hundredths place, and the digit 7 is in the thousandths place.

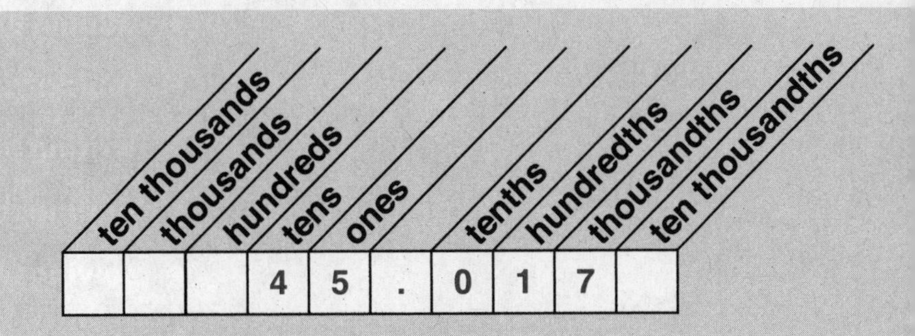


**Write each number in the place-value chart at the right. Then write each digit in the correct number group below.**

**1.** 6.4

6 ones

4 tenths

**2.** 1.73

_____ ones

_____ tenths

_____ hundredths

**3.** 10.28

_____ tens

_____ ones

_____ tenths

_____ hundredths

**4.** 0.955

_____ ones

_____ tenths

_____ hundredths

_____ thousandths

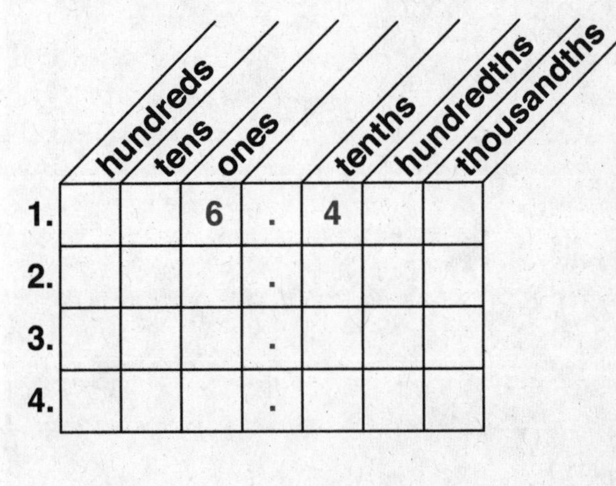


**Write the digit in the tenths place.**

**5.** 456.71 7   **6.** 0.9 _____   **7.** 2.071 _____   **8.** 9.3 _____   **9.** 0.18 _____   **10.** 26.428 _____

**Write the digit in the hundredths place.**

**11.** 0.306 0   **12.** 43.68 _____   **13.** 2.71 _____   **14.** 0.254 _____   **15.** 0.042 _____   **16.** 37.329 _____

**Check your answers on page 167.**

# Writing Decimals

To write a decimal in words, use the place name of the right-hand digit. Use the word *and* for the decimal point.

**17.6**
Write the whole number in words. Write *and* for the decimal point. The right-hand digit is in the tenths place.

**17.6 =**
**seventeen and six tenths**

**17.06**
The right-hand digit is in the hundredths place.

**17.06 =**
**seventeen and six hundredths**

**17.006**
The right-hand digit is in the thousandths place.

**17.006 =**
**seventeen and six thousandths**

**Put each decimal in the chart to the right. Then write each decimal in words below.**

1. 14.5 **fourteen and five tenths**
2. 8.225 ____________________
3. 30.2 ____________________
4. 7.04 ____________________
5. 508.01 ____________________
6. 0.63 ____________________
7. 4.9 ____________________
8. 40.108 ____________________

| | thousands | hundreds | tens | ones | . | tenths | hundredths | thousandths |
|---|---|---|---|---|---|---|---|---|
| 1. | | | 1 | 4 | . | 5 | | |
| 2. | | | | | . | | | |
| 3. | | | | | . | | | |
| 4. | | | | | . | | | |
| 5. | | | | | . | | | |
| 6. | | | | | . | | | |
| 7. | | | | | . | | | |
| 8. | | | | | . | | | |

**Write each number as a decimal.**

9. seventy and thirty-three hundredths

**70.33**

10. nine and fifty-two thousandths

____________________

11. twenty-six hundredths

____________________

12. four hundred nine thousandths

____________________

**Check your answers on page 167.**

# Zeros and Decimals

Zeros can be added to the right of a decimal at the end of a number without changing the value of that number.

$9 = 9.0$ $20 = $20.00 $.5 = .50 = .500$

Zeros can be dropped from the right of a decimal at the end of a number without changing the value of that number.

$14.0 = 14$ $3.00 = $3 $7.00 = 7.0 = 7$

Zeros must *not* be dropped from the middle of a decimal. When zeros are dropped from the middle of a decimal, the value of a number changes. The symbol ≠ means *is not equal to*.

$6.05 \neq 6.5$ $804.09 ≠ $84.9

**Decide if these decimals are equal:** 2.70 ☐ 2.7

**1.** If there is a zero at the end of the decimal, you can drop it.

**2.70̸ ☐ 2.7**

**2.** The decimals are equal. Write = in the box.

**2.70 [=] 2.7**

**Decide if the decimals are equal. Write = or ≠ in each box.**

**1.** 14 [≠] 104.00
**14 104.0̸0̸**
**14 ≠ 104**

**2.** $12 ☐ $12.00

**3.** 55.01 ☐ 55.1

**4.** 0.230 ☐ 0.203

**5.** 181.0 ☐ 181

**6.** 30.600 ☐ 30.6

**7.** 250.0 ☐ 2.5

**8.** $0.04 ☐ $0.40

**Fill in the blanks.**

**9.** Nineteen and zero tenths **is equal to** nineteen and zero hundredths.

**10.** Sixty and two hundredths ______________ sixty and twenty thousandths.

**11.** One and three tenths ______________ one and thirty thousandths.

**Check your answers on page 168.**

# Comparing Decimals

To compare two decimals, line up the decimal points. Then compare the digits starting from the left. The greater number is the number with the greater digit farthest to the left. Compare until you find two digits that are different. The symbol for *less than* is <. The symbol for *greater than* is >.

**Compare 23.57 and 23.75.**

23.57
23.75

The tens and ones digits are the same. The tenths digits are different.

5 tenths < 7 tenths, so
**23.57 < 23.75**

**Compare 0.3 and 0.071.**

0.3
0.071

The ones digits are the same. The tenths digits are different.

3 tenths > 0 tenths, so
**0.3 > 0.071**

**Compare 1.080 and 1.08.**

1.080
1.08

A right-hand zero does not change the value of a decimal.

80 thousandths = 8 hundredths, so **1.080 = 1.08**

**Decide if the decimals are equal. Write = or ≠ in each box.**

1. 0.03 [=] 0.030
   **0.03**
   **0.030**
2. 2.10 [ ] 1.20
3. 60.50 [ ] 60.5

**Compare the decimals. Write >, <, or =.**

4. 0.73 [<] 0.83
   **.73**
   **.83**
5. 0.9 [ ] 0.91
6. 0.40 [ ] 0.4
7. 0.11 [ ] 1.1
8. 1.362 [ ] 0.363
9. 25.06 [ ] 25.6
10. 9.09 [ ] 0.099
11. 1.033 [ ] 1.33

**Order each group of decimals from least to greatest. Write 1 in the box next to the decimal that is the least, and write 2 in the box next to the decimal that is the next least. Write 3 in the box next to the decimal that is the greatest.**

12. 7.1 [2]
    7.01 [1]
    7.11 [3]
13. 0.003 [ ]
    0.3 [ ]
    0.03 [ ]
14. 50.5 [ ]
    5.05 [ ]
    0.05 [ ]

**Check your answers on page 168.**

# Rounding Decimals to the Nearest Tenth

To round a decimal to the nearest tenth, look at the digit in the hundredths place. If it is less than 5, drop all the digits to the right of the tenths place.

0.12 rounds to 0.1
7.443 rounds to 7.4
26.008 rounds to 26.0

If the digit in the hundredths place is greater than or equal to 5, add 1 to the number in the tenths place. Drop all digits to the right of the tenths place.

9.59 rounds to 9.6
3.761 rounds to 3.8
0.084 rounds to 0.9

A number line can help you understand how to round decimals.

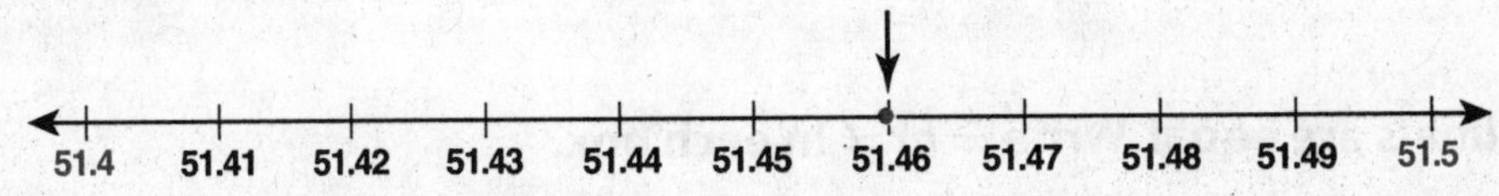


**Round 51.46 to the nearest tenth.**

**1.** 51.46 is between 51.4 and 51.5. Locate these numbers on the number line above.

Mark **51.46** on the number line.

**2.** 51.45 is halfway between 51.4 and 51.5.

Numbers less than 51.45 round down to 51.4.

Numbers greater than or equal to 51.45 round up to 51.5.

**3.** 51.46 is greater than 51.45. So, 51.46 rounds up to 51.5.

**51.46 → 51.5**

**Use the number line above to help you round each decimal to the nearest tenth.**

**1.** 51.45 **51.5** **2.** 51.42 ____ **3.** 51.44 ____ **4.** 51.47 ____ **5.** 51.41 ____ **6.** 51.48 ____

**Round each decimal to the nearest tenth.**

**7.** 0.42 **0.4** **8.** 23.601 ____ **9.** 3.152 ____ **10.** 184.01 ____ **11.** 17.629 ____ **12.** 8.18 ____

**Check your answers on page 168.**

# Rounding Decimals to the Nearest Hundredth

To round a decimal to the nearest hundredth, look at the digit in the thousandths place. If it is less than 5, drop all the digits to the right of the hundredths place.

3.522 rounds to 3.52
68.1406 rounds to 68.14
0.7932 rounds to 0.79

If the digit in the thousandths place is greater than or equal to 5, add 1 to the number in the hundredths place. Drop all digits to the right of the hundredths place.

3.1589 rounds to 3.16
0.267 rounds to 0.27
4.0081 rounds to 4.01

A number line can help you understand how to round decimals.

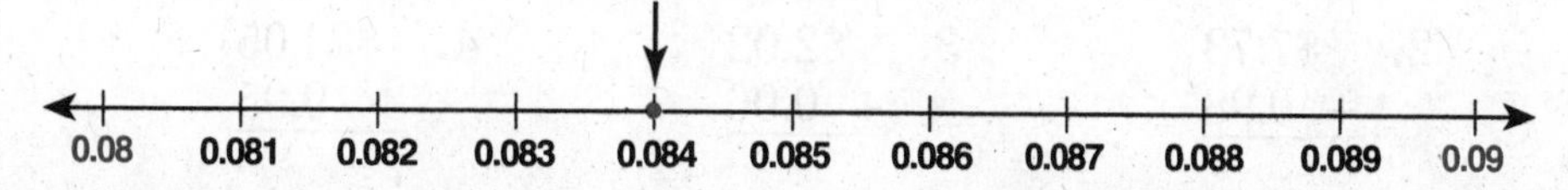


**Round 0.084 to the nearest hundredth.**

**1.** 0.084 is between 0.08 and 0.09. Locate these numbers on the number line above.

Mark **0.084** on the number line.

**2.** 0.085 is halfway between 0.08 and 0.09.

Numbers less than 0.085 round down to 0.08.

Numbers greater than or equal to 0.085 round up to 0.09.

**3.** 0.084 is less than 0.085. So, 0.084 rounds down to 0.08.

**0.084 → 0.08**

**Use the number line above to help you round each decimal to the nearest hundredth.**

**1.** 0.089 **0.09** **2.** 0.087 ____ **3.** 0.0892 ____ **4.** 0.085 ____ **5.** 0.083 ____ **6.** 0.086 ____

**Round each decimal to the nearest hundredth.**

**7.** 1.431 **1.43** **8.** 0.208 ____ **9.** 16.6152 ____ **10.** 50.002 ____ **11.** 38.374 ____ **12.** 19.426 ____

**Check your answers on page 168.**

# Adding Decimals

Add decimals as you do whole numbers. Be careful to line up the decimal point in the answer with the decimal points in the problem.

**Add:** 6.72 + 3.9

1. Set up the problem by lining up the decimal points in both numbers. Add a zero to make the same number of places.

```
 6.72
+3.90  ← add a zero
```

2. Add. Begin with the digits on the right. Rename. Put a decimal point in the answer.

```
  1
  6.72
 +3.90
 10.62
```

**Add. Rename if necessary.**

```
1.   1          2.  $7.73     3.  $2.02     4.  $34.05    5.  28.4
     8.7          + 0.98        + 0.06        +  9.95       +31.6
    +4.4
    13.1

6.  $0.27       7.  $300.40   8.  0.6       9.  2.1      10.  39.7
  + 0.75          +  51.85        3.0           3.0          515.22
                                 +1.05         +24.8        +  3.0
```

**11.** 3.1 + 2.96 =

```
  1
  3.10  ← add a zero
 +2.96
  6.06
```

**12.** $51.50 + $5 =

**13.** 88.4 + 21 =

**14.** 63.7 + 2.83 =

**Answer the word problems.**

**15.** Clarice bought a sweater for $25. The sales tax was $1.25. How much did Clarice spend for the sweater, including tax?

**16.** Last week Tonia worked 36.75 hours. This week she worked 41.5 hours. How many hours did Tonia work all together?

**Check your answers on page 169.**

# Subtracting Decimals

Subtract decimals as you do whole numbers. Be careful to line up the decimal points.

**Subtract:** 3.4 – 0.95

1. Set up the problem by lining up the decimal points to make the same number of places. Add a zero.

```
 3.40  ← add a zero
-0.95
```

2. Subtract. Begin with the digits on the right. Rename. Put a decimal point in the answer.

```
 2 1310
 3.40
-0.95
 2.45
```

**Subtract. Rename if necessary.**

1. 6.40 − 3.12 = 3.28 (renamed: 3 10)
2. $5.00 − 2.93
3. 1.7 − 0.6
4. 4.08 − 0.19
5. $12.15 − 3.06
6. 4.7 − 3.52
7. 8 − 2.75
8. 1.4 − 1.36
9. 2. − 0.91
10. 54.7 − 23.94

11. $5.17 – $0.48 =

```
  4 1017
 $5.17
-  0.48
 $4.69
```

12. 3 – 0.6 =

13. 7.2 – 5.52 =

14. $8 – $3.11 =

**Answer the word problems.**

15. Rasha was running a marathon, or 26.2 miles. She pulled a muscle and dropped out after 16.75 miles. How many miles of the marathon were not completed?

16. Max bought a package of blank CDs for $9.45, including tax. He paid for the CDs with a $20 bill. How much change did Max receive?

**Check your answers on page 169.**

# Multiplying Decimals by Whole Numbers

To multiply decimals by whole numbers, line up the digits. The number of decimal places in the answer is the same as the number of decimal places in the problem.

**Multiply:** \$5.47 × 9

**1.** Set up the problem.

```
 $5.47
×    9
```

**2.** Multiply. Rename.

```
   4 6
 $5.47
×    9
  4923
```

**3.** Put the decimal point in the answer.

```
   4 6
 $5.47 ← two decimal places
×    9 ← no decimal places
$49.23 ← two decimal places
```

**Multiply.**

**1.**
```
  1
 5.2
× 8
41.6
```

**2.**
```
 $2.64
×   10
```

**3.**
```
 $31.50
×     5
```

**4.**
```
 40.6
×   3
```

**5.**
```
 $0.71
×   23
```

**6.**
```
  8.5
×  12
```

**7.**
```
 $50.26
×    40
```

**8.**
```
 $1.82
×   19
```

**9.** 3.5 × 5 =

```
 2
 3.5
×  5
17.5
```

**10.** 0.02 × 9 =

**11.** \$4.75 × 31 =

**12.** 2.09 × 15 =

**Answer the word problems.**

**13.** How much would you spend for 2 pounds of dried apricots at \$3.69 per pound?

**14.** A factory shipped 20 boxes to a customer. Each box weighed 12.5 pounds. What was the total weight of the shipment?

**Check your answers on page 170.**

# Multiplying Decimals by Decimals

Multiply a decimal by another decimal the way you do whole numbers. The number of decimal places in the answer is the same as the total number of decimal places in the problem.

**Multiply:** 3.4 × 2.6

1. Set up the problem.

```
 3.4
×2.6
```

2. Multiply. Rename.

```
   2
  3.4
 ×2.6
  204
 +68
  884
```

3. Count the number of decimal places in the problem. Put the decimal point in the answer to show the total number of decimal places.

```
   2
  3.4  ← one decimal place
 ×2.6  ← one decimal place
  204
 +68
 8.84  ← two decimal places
```

**Multiply. Round money amounts to the nearest cent.**

1.
```
     1
   0.19
 × 0.02
 0.0038
```

2.
```
  0.26
 × 0.2
```

3.
```
  0.5
 ×0.7
```

4.
```
 $2.43
× 0.08
```

5.
```
  4.71
 ×0.05
```

6.
```
 $20.00
 ×  4.5
```

7.
```
  7.5
 ×6.3
```

8.
```
  3.31
 × 8.3
```

9. 0.2 × 8.1 =
```
   1
  0.2
 ×8.1
  0 2
×1 6
 1.6 2
```

10. 0.58 × 0.9 =

11. $4.03 × 6.1 =

12. 1.71 × 5.2 =

**Answer the word problems.**

13. Ursula earns $16.80 for each hour of overtime. Last week she worked 6.5 hours overtime. How much overtime pay did Ursula earn last week?

14. Hector walked at an average speed of 3.6 miles per hour for 0.75 hours. How many miles did Hector walk?

**Check your answers on page 170.**

# Dividing Decimals by Whole Numbers

Divide decimals by whole numbers as you do whole numbers by whole numbers. Be sure to correctly place the decimal point in the quotient.

**Divide:** 2.6 ÷ 4

**1.** Set up the problem. Put a decimal point in the answer above the decimal point in the problem. Since you can't divide 2 by 4, put a 0 in the quotient as a placeholder.

$$\begin{array}{r} 0. \\ 4\overline{)2.6} \end{array}$$

**2.** Divide. Add zeros to the dividend as needed. Keep dividing until the remainder is 0.

$$\begin{array}{r} 0.65 \\ 4\overline{)2.60} \\ -2\,4 \\ \hline 20 \\ -20 \\ \hline 0 \end{array}$$

**Divide until the remainder is zero.**

**1.** $6\overline{)5.04}$

$$\begin{array}{r} 0.84 \\ 6\overline{)5.04} \\ -48 \\ \hline 24 \\ -24 \\ \hline 0 \end{array}$$

**2.** $15\overline{)2.4}$

**3.** $8\overline{)\$12}$

**4.** $4\overline{)0.25}$

**5.** 3.15 ÷ 5 =

$$\begin{array}{r} 0.63 \\ 5\overline{)3.15} \\ -3\,0 \\ \hline 15 \\ -15 \\ \hline 0 \end{array}$$

**6.** 0.918 ÷ 3 =

**7.** 240.6 ÷ 2 =

**8.** 13.09 ÷ 7 =

**9.** 61.35 ÷ 15 =

**10.** 2.88 ÷ 36 =

**11.** 596.8 ÷ 80 =

**12.** 17.5 ÷ 25 =

**Answer the word problems.**

**13.** A filmmaker used 68.46 feet of film to make 3 movies of equal length. How many feet of film were used for each movie?

**14.** Ann bought 4 quarts of motor oil for $6.36. If each quart cost the same, what was the price of each quart?

**Check your answers on page 171.**

# Dividing Decimals by Whole Numbers with Rounding

If you want the answer to a division problem to have one decimal place, divide to two decimal places and then round. If you want an answer to have two places, divide to three places and then round.

**Divide:** 2.9 ÷ 6. Round to the nearest hundredth.

**1.** Set up the problem. Put a decimal point in the answer.

$6\overline{)2.9}$

**2.** Add zeros to divide to three decimal places.

$$\begin{array}{r} 0.483 \\ 6\overline{)2.900} \\ -2\,4 \\ \hline 50 \\ -48 \\ \hline 20 \\ -18 \\ \hline 2 \end{array}$$

**3.** Round to the nearest hundredth.

0.483 rounds to **0.48**

**Divide. Round to the nearest tenth.**

**1.** $$\begin{array}{r} 4.28 \\ 8\overline{)34.26} \\ -32 \\ \hline 22 \\ -16 \\ \hline 66 \\ -64 \\ \hline 2 \end{array}$$

**4.28 rounds to 4.3**

**2.** $5\overline{)47.16}$

**3.** 9.1 ÷ 4 =

**4.** 62.64 ÷ 18 =

**Divide. Round to the nearest hundredth or cent.**

**5.** $$\begin{array}{r} 0.072 \\ 7\overline{)0.509} \\ -49 \\ \hline 19 \\ -14 \\ \hline 5 \end{array}$$

**0.072 rounds to 0.07**

**6.** $50\overline{)\$2.60}$

**7.** 2.85 ÷ 20 =

**8.** $4.02 ÷ 45 =

**Answer the word problems.**

**9.** Jason bought 2 dozen golf balls for $47.70. What was the rounded cost of each golf ball? (1 dozen = 12)

**10.** A $34.25 bill is shared equally among 3 roommates. What is each roommate's rounded cost?

**Check your answers on page 171.**

# Dividing Decimals by Decimals

To divide a decimal by another decimal, move the decimal points in both numbers the same number of places to the right. You may need to add zeros to the number you are dividing into.

**Divide:** $29.9 \div 4.6$

**1.** Set up the problem.

$$4.6\overline{)29.9}$$

**2.** Make the divisor a whole number, 46. Move both decimal points one place to the right. Put a decimal point in the answer.

$$4.6\overline{)29.9} \rightarrow 46\overline{)299.}$$

**3.** Divide until the remainder is zero. You need to add a zero to 299.

$$\begin{array}{r} 6.5 \\ 46\overline{)299.0} \\ -276 \\ \hline 23.0 \\ -23.0 \\ \hline 0 \end{array}$$

**Divide until the remainder is zero.**

**1.** $\begin{array}{r} 53 \\ 0.2\overline{)10.6} \\ -10 \\ \hline 06 \\ -6 \\ \hline 0 \end{array}$

**2.** $0.3\overline{)\$5.40}$

**3.** $0.232 \div 0.29 =$

**4.** $0.92 \div 2.3 =$

**Divide. Round to the nearest hundredth or cent.**

**5.** $\begin{array}{r} 6.182 \\ 0.4\overline{)2.4730} \\ -24 \\ \hline 07 \\ -4 \\ \hline 33 \\ -32 \\ \hline 10 \\ -8 \\ \hline 2 \end{array}$

6.182 rounds to 6.18

**6.** $1.3\overline{)\$6.58}$

**7.** $\$6.50 \div 5.4 =$

**8.** $0.404 \div 0.25 =$

**Answer the word problems.**

**9.** Kirk bought 2.5 pounds of potato salad for $4.50. What was the cost per pound?

**10.** A gold chain is 7.5 inches long and costs $135. To the nearest cent, what is the cost of the chain per inch?

**Check your answers on page 172.**

# Rounding and Estimating with Decimals

To estimate answers to division problems, round one or both numbers until you can solve the problem using a basic division fact.

**Divide: $0.38\overline{)1.967}$**

1. Round the number you are dividing by to a number that is easy to divide by. Round 0.38 to 0.4.
   $0.38\overline{)1.967} = 0.4\overline{)1.967}$

2. Round the number you are dividing into to a number that is easy to divide into. Round 1.967 to 2.
   $0.4\overline{)1.967} = 0.4\overline{)2}$

3. Add a zero. Move the decimal points and divide to complete the estimate.
   $0.4\overline{)2.0} = 4\overline{)20}$ = 5; $20 - 20 = 0$

**Round both numbers to estimate each answer.**

1. $0.029\overline{)6.05}$
   $0.03\overline{)6.00} = 3\overline{)600}$ = 200; $600 - 600 = 0$
2. $3.2\overline{)15.168}$
3. $0.88\overline{)8.27}$
4. $29\overline{)1.213}$
5. $1.6\overline{)0.48}$
6. $2.08\overline{)0.019}$
7. $12.3\overline{)3.602}$
8. $56.85\overline{)0.481}$
9. $1.923 \div 4.03 =$
   $4\overline{)2.00}$ = 0.5; $2.00 - 2\,0 = 0$
10. $0.36 \div 0.887 =$
11. $0.048 \div 2.09 =$
12. $3.1 \div 14.94 =$

**Answer the word problems.**

13. Amy paid \$108.40 for 5 gallons of paint. Estimate the cost of one gallon of paint.

14. In March, Rick's dog ate 32.5 pounds of dog food. Estimate the number of pounds of food Rick's dog ate each day. (Hint: March has 31 days.)

Check your answers on page 172.

## Using a Calculator with Decimals

You can use a calculator to work problems with decimals. Always clear your calculator with the AC key before you begin. Check each number after you enter it. Round answers as directed.

**Strategy** Learn how to add and subtract decimals on the Casio *fx-260* calculator. Follow these steps.

**Add: 46.3 + 22.9 – 33.6**

| | Press the key. | Read the display. |
|---|---|---|
| **1.** Clear the calculator. | → AC | → 0. |
| **2.** Enter the first number. | → 4 6 . 3 | → 46.3 |
| **3.** Press the add key. | → + | → 46.3 |
| **4.** Enter the second number. | → 2 2 . 9 | → 22.9 |
| **5.** Press the subtract key. | → – | → 69.2 |
| **6.** Enter the third number. | → 3 3 . 6 | → 33.6 |
| **7.** Press the equals key. Read the amount. | → = | → 35.6 |

**Strategy** Learn how to multiply and divide decimals on the calculator. Follow these steps.

**Divide: $589.67 ÷ 12 × 5**

| | Press the key. | Read the display. |
|---|---|---|
| **1.** Clear the calculator. | → AC | → 0. |
| **2.** Enter the dividend. | → 5 8 9 . 6 7 | → 589.67 |
| **3.** Press the divide key. | → ÷ | → 589.67 |
| **4.** Enter the divisor. | → 1 2 | → 12. |
| **5.** Press the multiply key. | → × | → 49.13916667 |
| **6.** Enter the number. | → 5 | → 5. |
| **7.** Press the equals key. Round to the nearest cent. $245.70 | → = | → 245.6958333 |

**Exercise 1: Use a calculator to add or subtract.**

1. $\begin{array}{r} 0.45 \\ +3.70 \\ \hline \end{array}$

2. $\begin{array}{r} \$3.80 \\ -\ 0.97 \\ \hline \end{array}$

3. $\begin{array}{r} \$42.67 \\ -\ 3.89 \\ \hline \end{array}$

4. $\begin{array}{r} 12.78 \\ +\ 6.5 \\ \hline \end{array}$

5. 0.42 – 0.08 =

6. \$0.50 + \$0.72 =

7. 12.06 – 2.7 =

8. \$2.31 + \$0.58 =

**Exercise 2: Use a calculator to multiply or divide. Round to the nearest hundredth or cent.**

9. 5.46 × 7 =

10. 0.6 × 1.8 =

11. 0.247 ÷ 5 =

12. \$0.95 ÷ 3.2 =

13. 0.45 × \$2.39 =

14. \$343.88 ÷ 6 =

15. 31 × \$2.79 =

16. 2.6 ÷ 0.41 =

**Exercise 3: Use a calculator to solve the word problems.**

17. Divide to find the average speed of a bicyclist who completes a 48.125-mile race in 1.75 hours.

18. A car rents for \$31.95 per day. What does it cost to rent the car for 7 days?

19. A hotel room is \$89.99 per night plus \$6.30 in tax. What is the total nightly cost for the room with tax?

20. Yolanda makes \$12.55 per hour. Before her last raise, she was paid \$11.60 per hour. How much was the raise?

**Check your answers on page 172.**

# Meaning of Percent

**Percent** means hundredths. When using percents, the whole is divided into 100 equal parts. Twenty-five percent (25%) means 25 hundredths, or 25 out of 100 parts. The sign % is read *percent.*

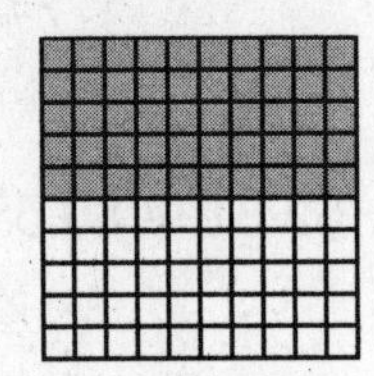

50%
fifty percent

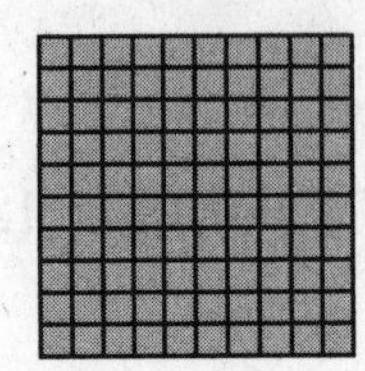

100%
one hundred percent

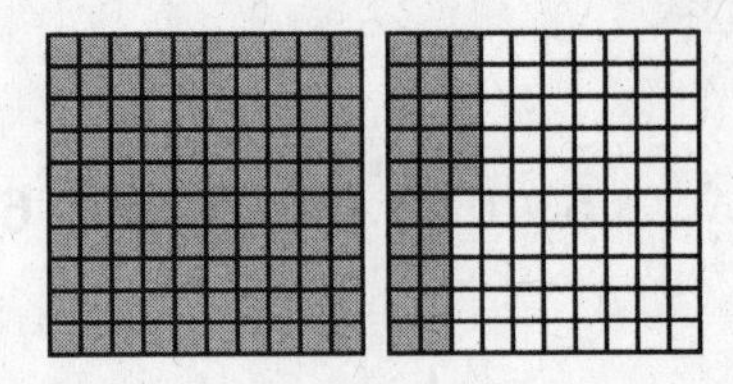

125%
one hundred twenty-five percent

**Write a percent for each figure.**

**1.** 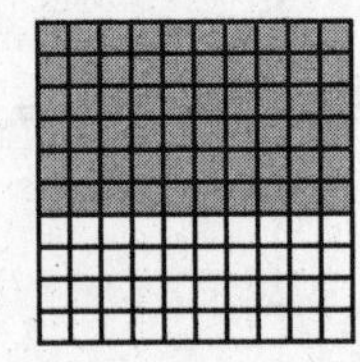

60%

**2.** 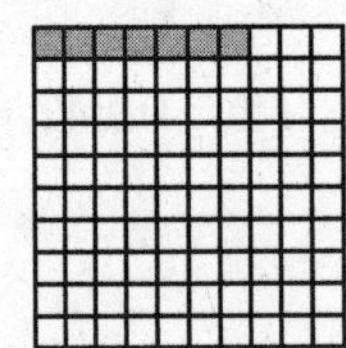

______

**3.** 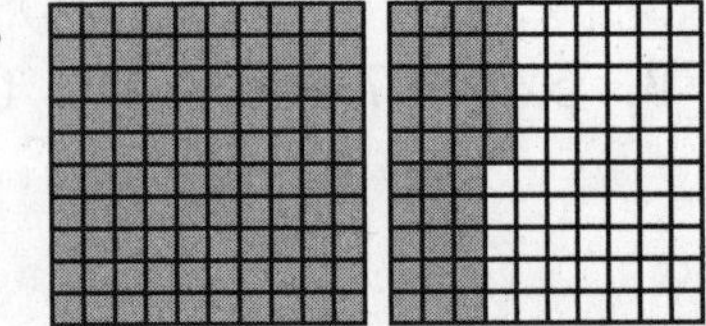

______

**4.** 

______

**5.** 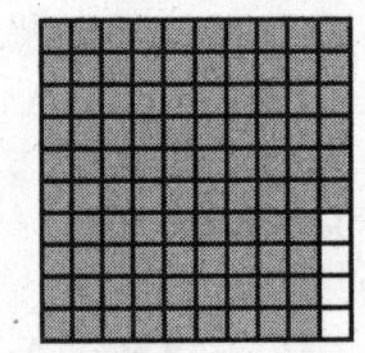

______

**6.** 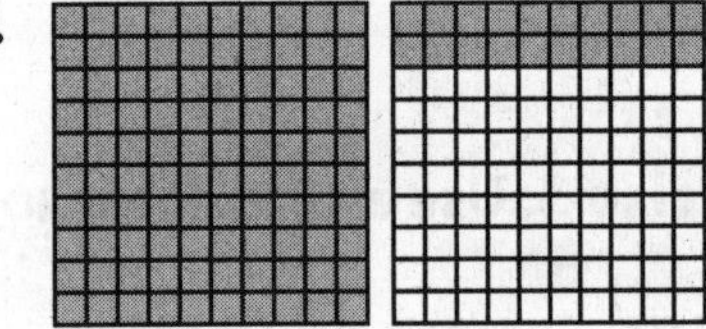

______

**Write a percent using the percent sign.**

**7.** one percent = 1%

**8.** twenty-five percent =

**9.** ten percent =

**10.** forty-five percent =

**11.** fifty-eight percent =

**12.** two hundred percent=

**13.** eighty percent =

**14.** seventy-five percent =

**15.** sixteen percent =

**16.** one hundred fifty percent =

**17.** eight and one-half percent =

**Check your answers on page 173.**

# Percents as Decimals

You can write any percent as a decimal and any decimal as a percent.

| **Change 225% to a decimal.** | **Change 9% to a decimal.** | **Change 0.4 to a percent.** |
|---|---|---|
| Write the number without the percent sign. Move the decimal point two places to the left. | For percents less than 10%, write a zero in front of the number. Then move the decimal point two places to the left. | Add a zero to the right of the decimal. Then move the decimal point two places to the right. Add a percent sign. |
| **225% = 225 = 2.25** | **9% = 09 = 0.09** | **0.4 = 0.40 = 40%** |

**Change each percent to a decimal.**

**1.** 50% = **50. = 0.5** **2.** 380% = **3.** 5% = **05. = 0.05** **4.** 73.2% =

**5.** 112% = **6.** 18% = **7.** 49.5% = **8.** 130% =

**9.** 74% = **10.** 7.1% = **11.** 345% = **12.** 93.5% =

**Change each decimal to a percent.**

**13.** 0.045 = **0.045 = 4.5%** **14.** 2.3 = **15.** 0.85 = **16.** 0.2 =

**17.** 0.506 = **18.** 0.07 = **19.** 0.55 = **20.** 1.85 =

**21.** 0.318 = **22.** 0.04 = **23.** 3.41 = **24.** 0.16 =

**Answer the word problems.**

**25.** A soccer team scored 15% more goals this season than last season. Write 15% as a decimal.

**26.** Mark used 0.08 of the gas in the gas tank. What percent of the gas in the tank did Mark use?

**Check your answers on page 173.**

# Finding the Part

There are three pieces to a percent problem: the *whole*, the *percent*, and the *part*. Sometimes in a percent problem, one of the three pieces is missing. To solve for the part, multiply the whole by the percent (part = whole × percent).

**1. Write the whole, the percent, and the part. 30% of 200 is 60**

**whole = 200**

**percent = 30%**

**part = 60**

**2. What is 30% of 200?**

**a.** Write the pieces.

**whole = 200**

**percent = 30%**

**part = 200 × 30%**

**b.** Change the percent to a decimal.

**30% = 030. = 0.3**

**c.** Multiply 200 by 0.3 to find the part.

**200 × 0.3 = 60**

**Write the whole, the percent, and the part.**

**1.** 20% of 40 is 8
**whole = 40**
**percent = 20%**
**part = 8**

**2.** 75% of 48 is 36

**3.** 30% of 150 is 45

**4.** 60% of 25 is 15

**5.** 25% of 200 is 50

**6.** 150% of 80 is 120

**Find the percent of each number.**

**7.** What is 10% of 50?
**whole = 50**
**percent = 10% = 0.1**
**part = 50 × 0.1 = 5**

**8.** What is 80% of 400?

**9.** What is 120% of 150?

**10.** 30% of 140 is what?

**11.** 15% of 280 is what?

**12.** 110% of 600 is what?

**Answer the word problems.**

**13.** Lucia saves 15% of her income. She earns $1,600 each month. How much does Lucia save each month?

**14.** At the Davisville Auto Mart, 30% of the cars are blue. If there are 120 cars on the lot, how many cars are blue?

**Check your answers on page 174.**

# Finding the Percent

In a percent problem, the missing piece may be the percent. To solve for the percent, first divide the part by the whole (percent = part ÷ whole). Then write your answer as a percent.

**What percent of 50 is 10?**
**part = 10**
**whole = 50**
**percent = 10 ÷ 50 = 0.2**

Change 0.2 to a percent.
**0.2 = 0.20 = 20%**

**10 is 20% of 50**

**10 is what percent of 40?**
**part = 10**
**whole = 40**
**percent = 10 ÷ 40 = 0.25**

Change 0.25 to a percent.
**0.25 = 0.25 = 25%**

**10 is 25% of 40**

**Write the part, the whole, and the percent equation.**

1. What percent of 20 is 5?
   **part = 5**
   **whole = 20**
   **percent = 5 ÷ 20**
2. What percent of 40 is 15?
3. 8 is what percent of 80?
4. 4 is what percent of 100?
5. What percent of 28 is 56?
6. 270 is what percent of 360?

**Find the percent.**

7. 30 is what percent of 60?
   **part = 30**
   **whole = 60**
   **percent = 30 ÷ 60 = 0.50 = 50%**
8. What percent of 4 is 7?
9. 54 is what percent of 18?
10. What percent of 32 is 8?
11. 108 is what percent of 144?
12. What percent of 50 is 30?

**Check your answers on page 174.**

# GED Skill Strategy

## Reading Circle Graphs

Circle graphs are often used to show how a whole group is divided up into different percents. The total of all the percents must add up to 100%. The graph below shows the results of a survey. Parents were asked if they thought their town needed another school.

**Strategy** Learn how to read a circle graph. Follow these steps.

**Do We Need Another School?**

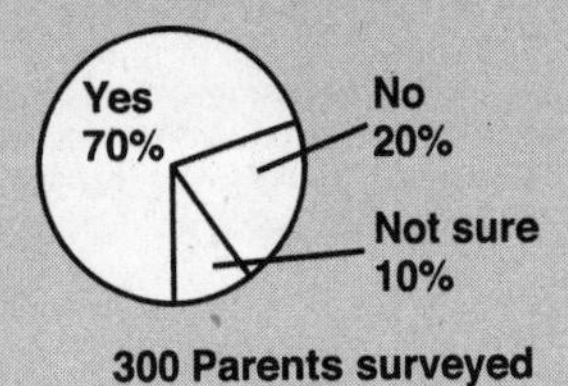


**1.** Read all the sections. Check to see they total 100%.

| | |
|---|---|
| **yes —** | **70%** |
| **no —** | **20%** |
| **not sure —** | **+10%** |
| | **100%** |

**2.** Decide what the graph is telling you. Most (70%) of the parents said they need another school.

**Exercise 1: Use the circle graphs below to answer the questions.**

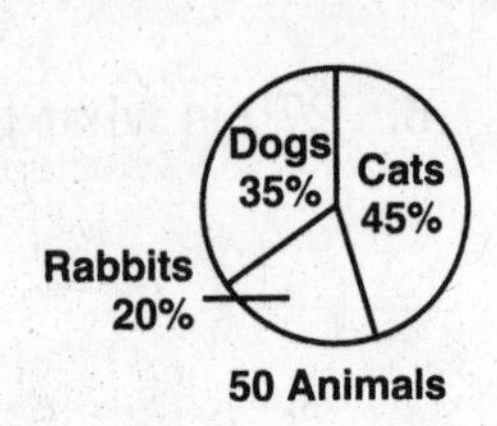


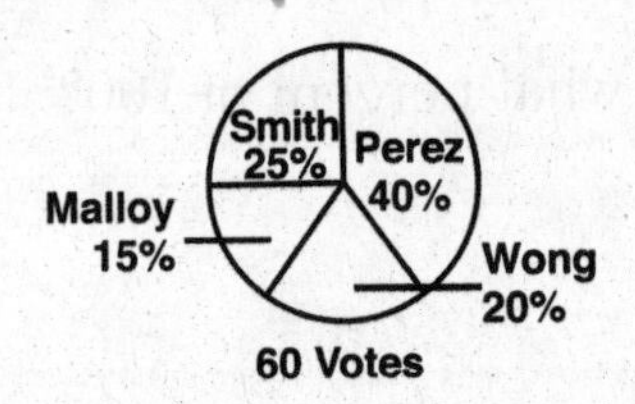


**1. a.** What percent are cats? 45%

**b.** What percent are dogs and cats? _____

**2. a.** What percent voted for Malloy? _____

**b.** What percent voted for Perez? _____

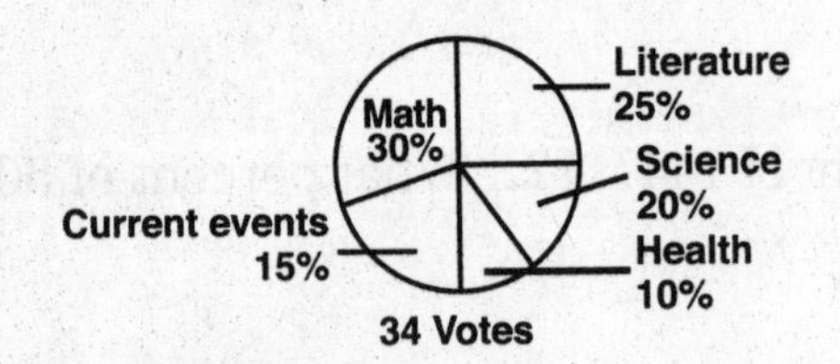


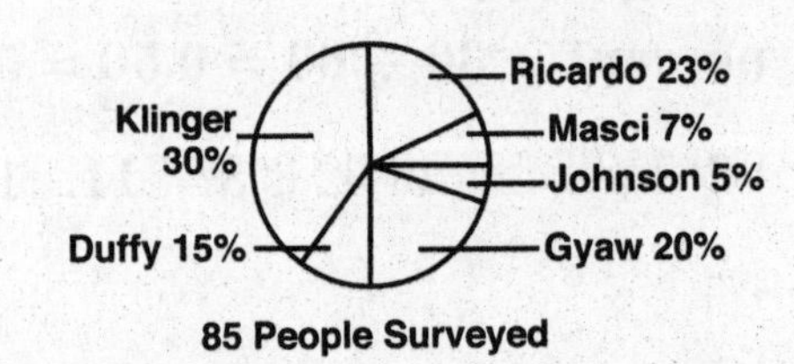


**3. a.** How many total votes? _____

**b.** What percent chose Math? _____

**4. a.** How many were surveyed? _____

**b.** What percent chose Masci? _____

**Exercise 2: Use the circle graphs below to answer the questions.**

**5.** Do We Need More Street Lights? 40 People Surveyed

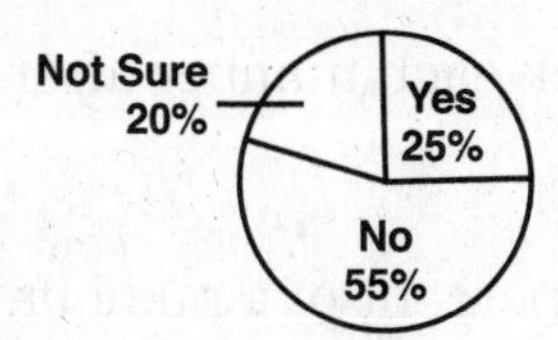


**a.** How many people were surveyed?

**b.** What percent voted yes? _____

**c.** What percent voted no? _____

**d.** What percent voted no or not sure? _____

**6.** Does Your Building Have Smoke Detectors? 60 People Surveyed

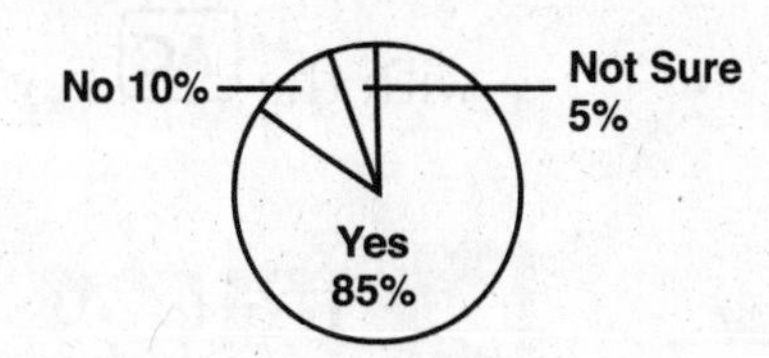


**a.** How many people were surveyed? _____

**b.** What percent said yes? _____

**c.** What percent said not sure? _____

**d.** What percent more said yes than no? _____

**7.** Budget for Monthly Income: $800

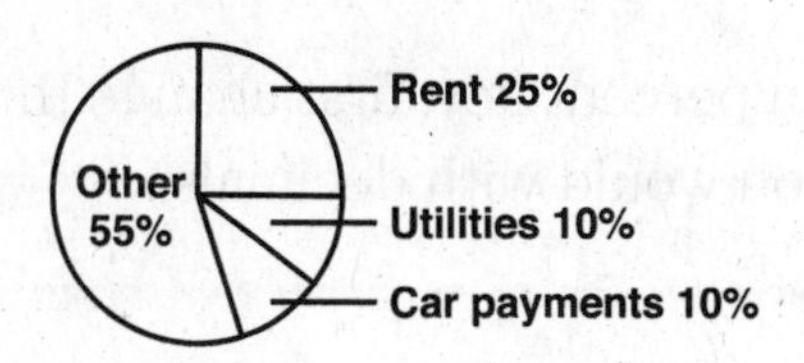


**a.** What is the total monthly income? _____

**b.** What percent is rent? _____

**c.** What percent is other? _____

**d.** What percent is rent and utilities? _____

**e.** What percent is utilities and car payments? _____

**8.** Don's Mini-Market Weekly Budget: $600

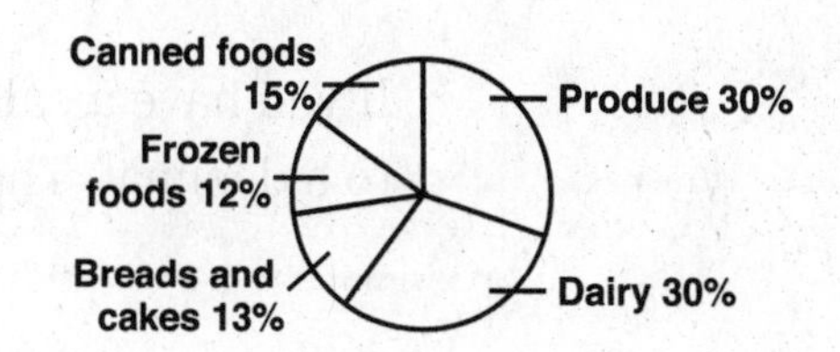


**a.** What is the weekly budget? _____

**b.** What percent is breads and cakes? _____

**c.** What percent is canned foods? _____

**d.** What percent is produce and dairy? _____

**e.** What is the difference in the percents for canned foods and frozen foods? _____

**Check your answers on page 174.**

## Using a Calculator with Percents

The Casio *fx-260* calculator has a percent key. To use the percent key, press the SHIFT key, and then the = key. Always clear your calculator with the AC key before you begin. Check each number after you enter it.

**Strategy** Learn how to find the percent of a number. Follow these steps.

**Find 10% of 60.**

| | Press the key. | Read the display. |
|---|---|---|
| 1. Clear the calculator. | → AC | → 0. |
| 2. Enter the <u>second</u> number. | → 6 0 | → 60. |
| 3. Press the multiply key. | → × | → 60. |
| 4. Enter the percent number. | → 1 0 | → 10. |
| 5. Press the SHIFT key and then the = key. | → SHIFT = | → 6. |

If you have a calculator without a percent key, first change the percent to a decimal. Then multiply as you would with decimals.

**Exercise 1: Use a calculator to solve.**

1. 15% of 140
2. 55% of 20
3. 5% of 30
4. 150% of 60
5. 75% of 120
6. 80% of 260
7. 150% of 64
8. 20% of 160
9. 40% of 360

**Exercise 2: Use a calculator to solve. Round answers to the nearest tenth when needed.**

**10.** 25% of 74

**11.** 125% of 65

**12.** 45% of 12

**13.** 6% of 32

**14.** 62% of 175

**15.** 11% of 88

**16.** 17.3% of 80

**17.** 9.5% of 36

**18.** 12% of 87.3

**Exercise 3: Use a calculator to solve the word problems.**

**19.** A furniture store is having a 15%-off sale. How much would you save on a sofa with a regular price of $800?

**20.** A company buys equipment that costs $150,000. It plans to borrow 85% of the cost. How much does the company plan to borrow?

**21.** The population of Fulton is 16,000. In the last election, 48% of the population voted. How many people voted?

**22.** A test has 40 questions. Andrew got 90% of the questions on the test correct. How many questions did Andrew answer correctly?

**Check your answers on page 175.**

# GED Skill Strategy

## Reading Bar Graphs

A bar graph is a way to show information in pictures. It includes bars of different lengths that stand for certain numbers. The bars in a bar graph can be drawn across or up and down.

These bar graphs show the amount of sales tax that is charged on purchases in various states.

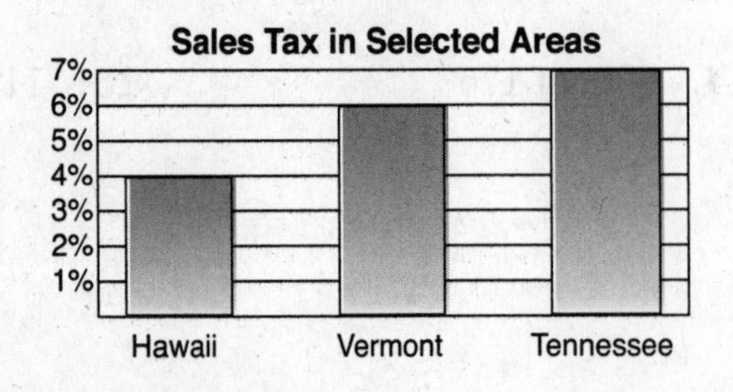


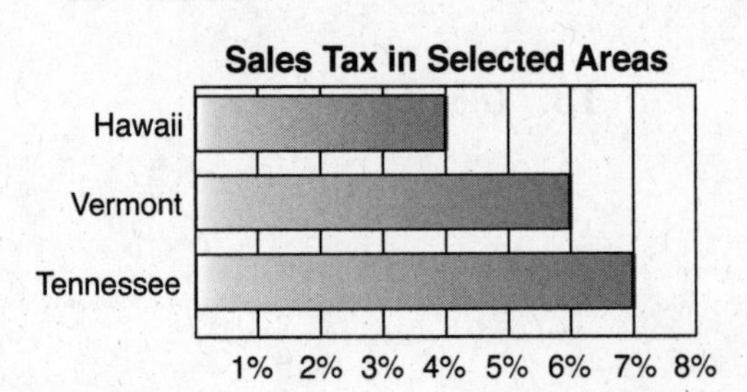

**Strategy** Learn how to use a bar graph.

How much sales tax will be charged for a $150 purchase in the state of Hawaii?

**1.** Sales tax in Hawaii is 4%.

**2. 4% sales tax; $150 purchase**

**3.** Change 4% to a decimal. Multiply.

**4% = 0.04  $150 × 0.04**

**4.** $$\begin{array}{r} \$150 \\ \times\, 0.04 \\ \hline \$6.00 \end{array}$$

**Exercise 1: Use the bar graphs above to solve each problem.**

**1.** How much sales tax will be charged for an $80 purchase in the state of Vermont?

**2.** How much sales tax will be charged for a $35 purchase in the state of Tennessee?

**3.** How much sales tax will be charged for a $60 purchase in the state of Hawaii?

**4.** How much sales tax will be charged for a $125 purchase in the state of Tennessee?

**Exercise 2: Use the bar graph to solve each problem.**

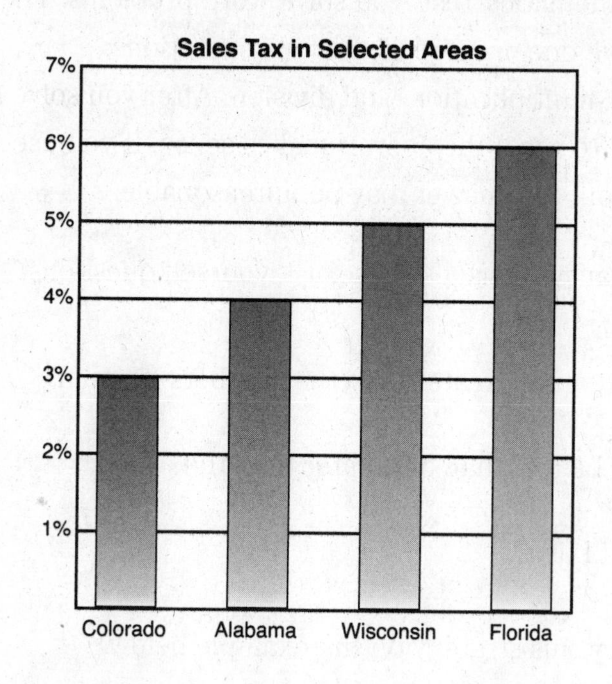


5. Which state shown on the graph charges the most sales tax?

6. Which state shown on the graph charges the least sales tax?

7. In Alabama, how much sales tax will be charged for a $500 purchase?

**Exercise 3: Use the bar graph to solve each problem.**

8. Which states shown on the graph charge the same amount of sales tax?

9. How much sales tax will be charged for a $125 purchase in Connecticut?

10. In Georgia, how much sales tax will be charged for a $36.99 purchase?

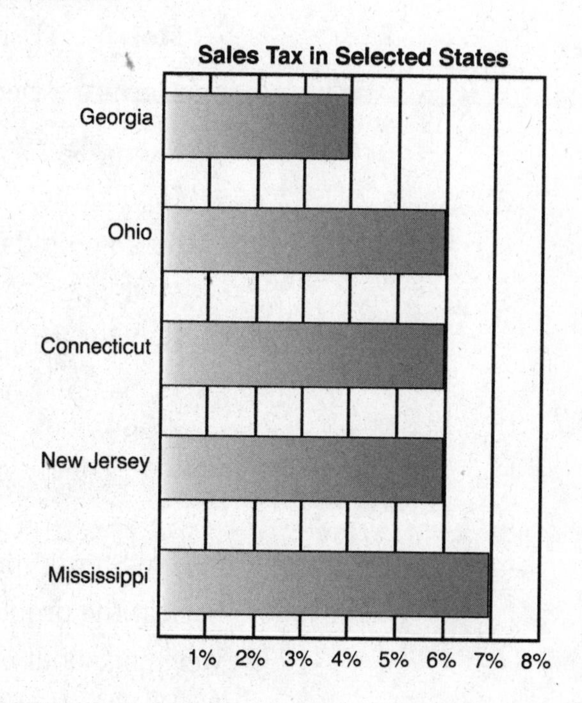


Check your answers on page 175.

## Solving to See if the Answer Makes Sense

On the GED Mathematics Test, you solve word problems. The problems often involve one or more of the four operations: addition, subtraction, multiplication, and division. After you solve a word problem, check to see if the answer makes sense. If you use the incorrect operation, the answer may be unreasonable.

To determine if your answer makes sense, ask yourself questions such as the following:

- Is the answer less than or greater than the numbers given in the problem?
- Should the answer be less than or greater than the given numbers?
- What operation did I use?

**Strategy** Try this strategy on the example below.

Use these steps.

**Step 1** Read the numbers in the answer choices.

**Step 2** Look back at the numbers in the problem.

**Step 3** Estimate the answer.

**Step 4** Decide if the answer is reasonable.

**Example**

A pallet holds 6 bags of fertilizer. Each bag weighs 48.25 pounds. How much total weight is on the pallet?

(1) 42.25
(2) 289.5
(3) 54.25
(4) 8

In Step 1 you read the answer choices. In Step 2 you looked back at the problem. In Step 3 you estimated the answer. Six bags weighing about 50 pounds gives an answer close to 300 pounds. In Step 4 you decided which answer is reasonable. Choice **(2)**, 289.5 pounds, is reasonable.

## Practice

**Practice the strategy. Use the steps you learned. Solve the problems.**

**1.** How much longer is song 1 than song 2?

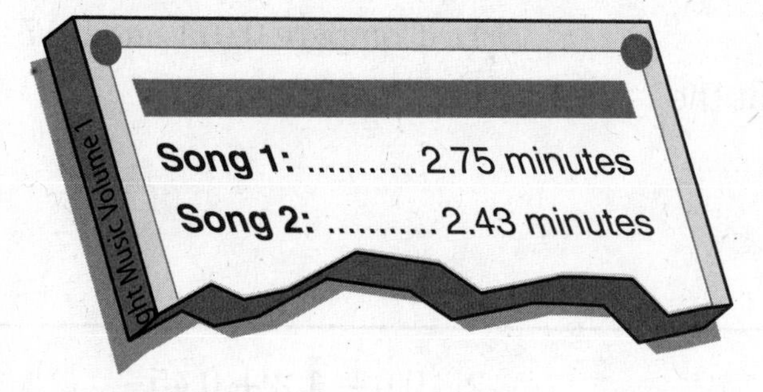


(1) 0.32 minute
(2) 5.18 minutes
(3) 6.68 minute
(4) 1.13 minutes

**2.** Below is Alicia's pay stub. What is her hourly rate?

Ruiz, Alicia Week ending 5/20

Hours Worked: 36
Gross Pay: $460.80

(1) $496.80
(2) $120.80
(3) $12.80
(4) $424.80

**3.** Manuel bought 16.5 feet of canvas. He used 4 feet to cover a chair. How many feet of canvas does Manuel have left?

(1) 20.5
(2) 12.5
(3) 66
(4) 4.13

**4.** How much would you pay for 0.91 pound of fruit salad if the salad costs $1.99 per pound?

(1) $1.08
(2) $1.81
(3) $2.90
(4) $2.19

**5.** Jan's dosage of medicine is 7.5 milliliters, 3 times a day. What is the total amount of milliliters of medicine she takes in a day?

(1) 10.5
(2) 4.5
(3) 2.5
(4) 22.5

**6.** Before starting a new exercise program, Tim weighed 175.2 pounds. After a month, he weighed 170.5 pounds. How many pounds did Tim lose?

(1) 4.7
(2) 14.7
(3) 4.07
(4) 0.47

**Check your answers on page 175.**

## Unit 4 Wrap-up

Below are examples of the skills you learned in this unit. Read the examples and solve the problems. Then check your answers.

**Examples.**

1. Compare 2.4 and 2.089

   **2 . 4**
   **2 . 0 8 9**

   Line up the decimal points. Start at the left. The ones digits are the same. 4 tenths > 0 tenths, so

   **2.4 > 2.089**

2. 7.9 + 8.27

   $$\begin{array}{r} 7.90 \\ +\ 8.27 \end{array} \leftarrow \text{add a zero} \qquad \begin{array}{r} \scriptstyle 1 \\ 7.90 \\ +\ 8.27 \\ \hline 16.17 \end{array}$$

3. 7.3 − 1.06

   $$\begin{array}{r} 7.30 \\ -\ 1.06 \end{array} \leftarrow \text{add a zero} \qquad \begin{array}{r} \scriptstyle 210 \\ 7.\not{3}\not{0} \\ -\ 1.06 \\ \hline 6.24 \end{array}$$

4. 32.4 × 0.05

   $$\begin{array}{r} 32.4 \\ \times\ 0.05 \end{array} \qquad \begin{array}{r} \scriptstyle 12 \\ 32.4 \\ \times\ 0.05 \\ \hline 1620 \end{array} \qquad \begin{array}{r} \scriptstyle 12 \\ 32.4 \\ \times\ 0.05 \\ \hline 1.620 \end{array} \begin{array}{l} \\ \leftarrow \text{one decimal place} \\ \leftarrow \text{two decimal places} \\ \leftarrow \text{three decimal places} \end{array}$$

5. 0.112 ÷ 1.4

   $$1.4\overline{)0.112} \qquad \begin{array}{r} 0.0 \\ 14\overline{)01.12} \end{array} \qquad \begin{array}{r} 0.08 \\ 14\overline{)01.12} \\ -1\ 12 \\ \hline 0 \end{array}$$

6. 60% of 200 is what?

   **whole = 200**
   **percent = 60%**
   **part = 200 × 60%**

   **60% = 0.60 = 0.6**
   **200 × 0.6 = 120**

**Problems.**

1. Compare 0.071 and 0.17
2. 0.6 + 1.2 + 0.45
3. 15.13 − 3.7
4. 1.7 × 0.32
5. 4.284 ÷ 0.7
6. 30% of 240 is what?

**Check your answers on page 176.**

# Unit 4 Practice

**Compare the decimals. Write >, <, or =.**

**1.** 1.3 ☐ 3.1  **2.** 2.10 ☐ 2.1  **3.** 0.6 ☐ 0.06  **4.** 1.91 ☐ 1.19

**Round each decimal to the nearest hundredth or cent.**

**5.** 2.345  **6.** 0.069  **7.** 12.008  **8.** $1.595  **9.** $.403  **10.** $8.097

**Add, subtract, multiply, or divide. Round division answers to the nearest hundredth.**

**11.** $0.61 + 1.8 =$  **12.** $3.2 + 0.5 + 1.09$  **13.** $0.2 \times 0.08 =$  **14.** $3.45 \times 1.9 =$

**15.** $7 \times 2.06 =$  **16.** $0.72 \times 30 =$  **17.** $1.2 \times 0.05 =$  **18.** $7.1 \times 11.4 =$

**19.** $4\overline{)9.106}$  **20.** $0.6\overline{)0.2485}$  **21.** $0.315 \div 17 =$  **22.** $0.1 \div 0.32 =$

**Answer the word problems.**

**23.** Rick saved $175. He spent 80% of his savings on a new bike. How much did Rick spend on the bike?

**24.** A bookstore has a budget of $480 for ads. The bookstore spent $288 on one ad. What percent of the budget was spent for the ad?

**Check your answers on page 176.**

# GED Test Practice

**Read each problem carefully. Circle the number of the correct answer.**

**1.** Reggie's time in the 200-meter run was twenty-four and eight-tenths seconds. How is this time written as a decimal?

(1) 248
(2) 24.8
(3) 24.08
(4) 2.48

**2.** The table shows the amount of fat in 3 ounces of different meats. List the meats in order based on the least to greatest amount of fat.

| Sirloin | Chicken | Pork |
|---|---|---|
| 6.1g | 3.6g | 7.2g |

(1) sirloin, chicken, pork
(2) chicken, pork, sirloin
(3) pork, sirloin, chicken
(4) chicken, sirloin, pork

**3.** How much change from $50 will a customer get for a purchase of $48.36?

(1) $1.64
(2) $1.74
(3) $2.64
(4) $98.36

**4.** Susana walked 3.5 miles in the morning and 2.8 miles in the afternoon. How many miles did she walk all together?

(1) 0.7
(2) 5.3
(3) 6.3
(4) 6.7

**5.** A piece of lumber is 188.25 inches long. If it is cut into 3 equal pieces, what will the length of each piece be?

(1) 564.75 inches
(2) 191.25 inches
(3) 185.25 inches
(4) 62.75 inches

**6.** A package contains 9 juice boxes. Each juice box holds the number of fluid ounces shown below. How many fluid ounces of juice are there all together in the package?

(1) 67.50
(2) 60.75
(3) 54.75
(4) 15.75

**7.** Jade had 11 gallons of gas in her car's tank. She drove 283.5 miles before the tank was just about empty. About how many miles per gallon of gas did Jade's car get?

(1) about 2
(2) about 11
(3) about 30
(4) about 250

**8.** At the Monroe Company, there are 140 employees. Of the 140 employees, 35% have a college degree. How many employees have a college degree?

(1) 35
(2) 49
(3) 56
(4) 105

**9.** Out of 50 bulbs, 3 were broken. What percent was broken?

(1) 6%
(2) 16%
(3) 47%
(4) 60%

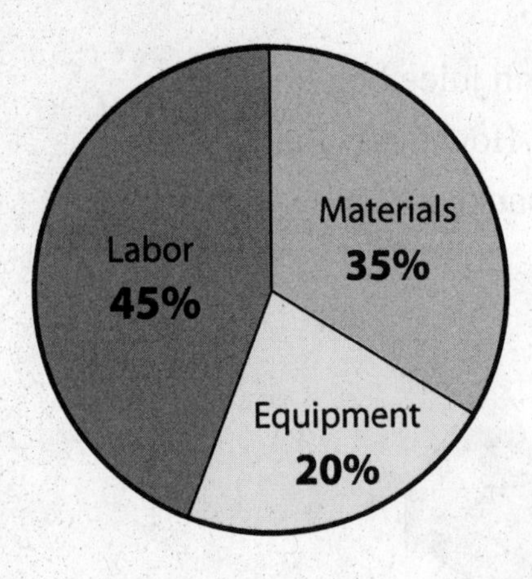


**10.** Look at the circle graph to the left. What percent of expenses are labor and materials?

(1) 10%
(2) 65%
(3) 80%
(4) 60%

**Check your answers on page 176.**

## Unit 4 Skill Check-Up Chart

Check your answers. In the first column, circle the numbers of any questions that you missed. Then look across the rows to see the skills you need to review and the pages where you can find each skill.

| Question | Skill | Page |
|---|---|---|
| 1 | Writing Decimals | 79 |
| 2 | Comparing Decimals | 81 |
| 3 | Subtracting Decimals | 85 |
| 4 | Adding Decimals | 84 |
| 5 | Dividing Decimals by Whole Numbers | 88 |
| 6 | Multiplying Decimals by Whole Numbers | 86 |
| 7 | Rounding and Estimating with Decimals | 91 |
| 8 | Finding the Part | 96 |
| 9 | Finding the Percent | 97 |
| 10 | Circle Graphs | 97–99 |

# Unit 5 Fractions, Ratios, and Proportions

mixed number

equal ratio

fraction

proportion

**In this unit you will learn about**

- the meaning of fractions
- equivalent fractions
- improper fractions and mixed numbers
- adding and subtracting fractions
- multiplying and dividing fractions
- reducing ratios to lowest terms
- solving proportions
- equivalent fractions, decimals, and percents

Fractions are used to name part of a whole or part of a group. You may use fractions when following a recipe, weighing ingredients at the grocery store, or taking a measurement. Write about a time you have used fractions in your life.

______________________________________________

______________________________________________

Ratios and proportions are used to compare two numbers. You use a ratio to compare the number of children to the number of adults on a field trip, or the number of cars to the number of trucks.

Write a ratio you have seen, such as the amount of water to the amount of juice mix needed to make a quart of juice.

______________________________________________

______________________________________________

# Meaning of Fractions

A fraction names part of a whole or part of a group. It is written as one number over another number. The **numerator**, or top number, tells how many parts of the whole or group are being considered. The **denominator**, or bottom number, tells how many equal parts are in the whole or group. The line that separates the numerator and denominator is called a **fraction bar**.

| | | | | |
|---|---|---|---|---|
| numerator | → | 3 | ← | number of parts being considered |
| fraction bar | → | — | ← | separates the numerator and denominator |
| denominator | → | 4 | ← | number of equal parts in the whole |

**Write a fraction for the parts that are shaded.**

**1.** Count the number of equal parts in the whole. This number is the denominator. The square is divided into 4 equal parts. Write 4 as the denominator.

$\overline{4}$

**2.** Count the number of shaded parts. This number is the numerator. The square has 3 shaded parts. Write 3 as the numerator.

$\frac{3}{4}$

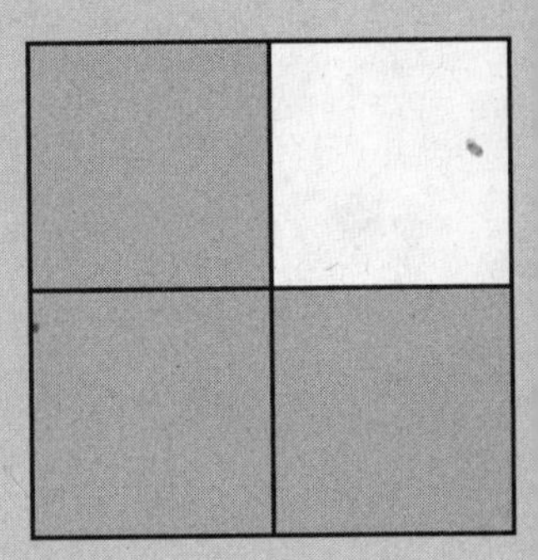

**Write a fraction for each shaded part.**

**1.** 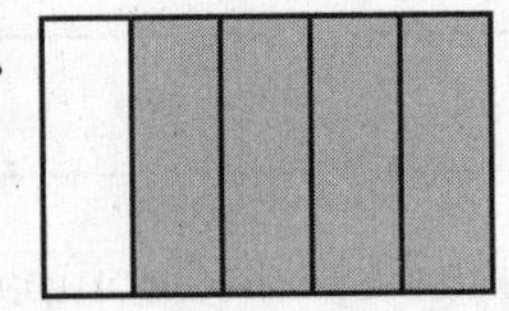

$\frac{4}{5}$

**2.** 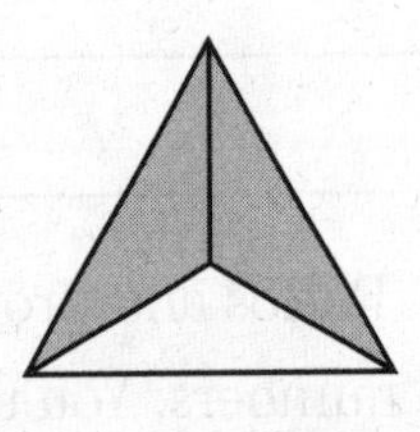

________

**3.** 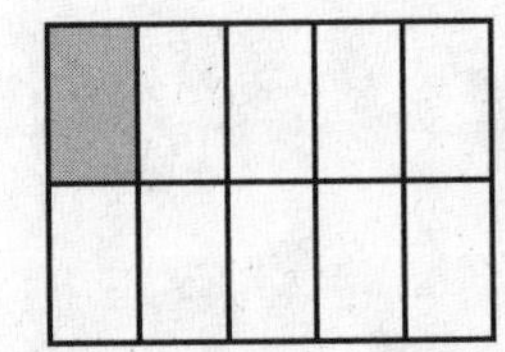

________

**4.** 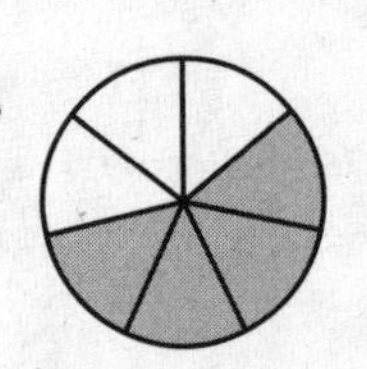

________

**Shade each figure to show the fraction.**

**5.** $\frac{3}{8}$

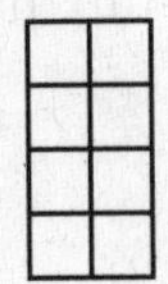

**6.** $\frac{5}{6}$

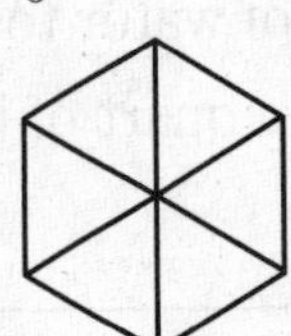

**7.** $\frac{1}{2}$

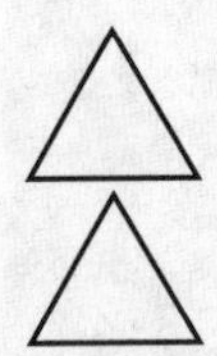

**8.** $\frac{3}{10}$

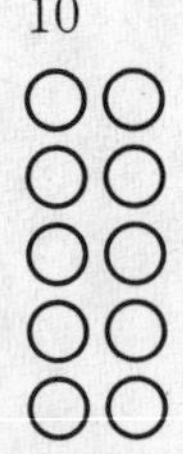

**Check your answers on page 177.**

# Equivalent Fractions

**Equivalent fractions** are different fractions that represent the same amount.

**What equivalent fractions are shown by these figures?**

**1.** Each of the large squares is one whole.

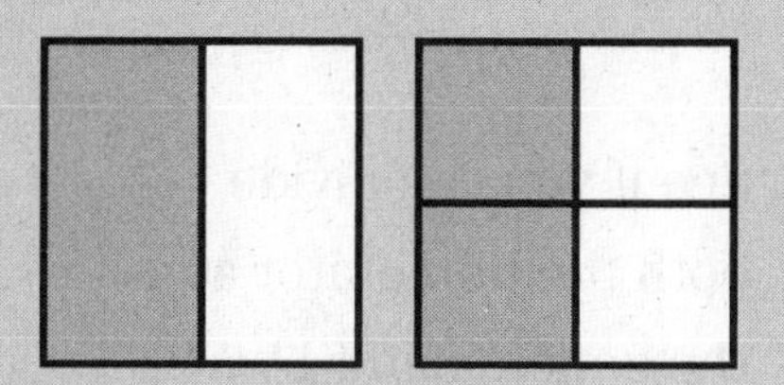

**2.** In the left square, 1 of the 2 equal parts ($\frac{1}{2}$) is shaded. In the right square, $\frac{2}{4}$ is shaded.

**3.** The same area is shaded in each square. The fractions $\frac{1}{2}$ and $\frac{2}{4}$ represent the same amount. They are equivalent fractions.

$$\frac{1}{2} = \frac{2}{4}$$

**What equivalent fractions are shown by these rulers?**

**1.** Each of these rulers is one inch long.

$\frac{3}{4}$ inch

Ruler A

**2.** The line on Ruler *A* measures $\frac{3}{4}$ inch. The line on Ruler *B* measures $\frac{12}{16}$ inch.

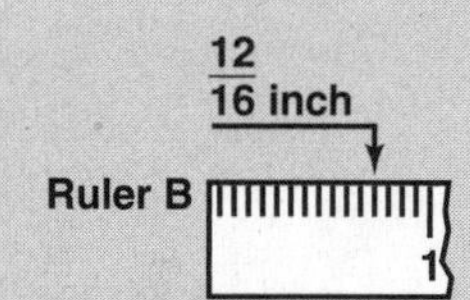


**3.** The lines are the same length. The fractions $\frac{3}{4}$ and $\frac{12}{16}$ show the same amount. They are equivalent fractions.

$$\frac{3}{4} = \frac{12}{16}$$

**Write equivalent fractions for each figure.**

**1.**

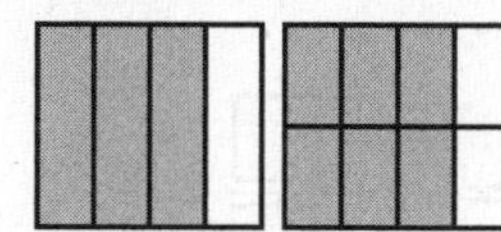

$\frac{3}{4} = \frac{6}{8}$

**2.**

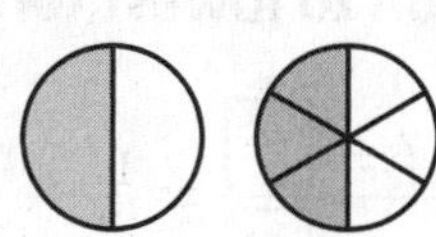

__________

**3.**

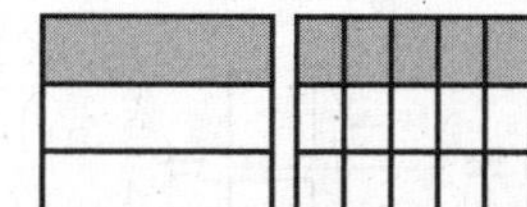

__________

**Write equivalent fractions for each pair of rulers.**

**4.** Line *A* and Line *B*

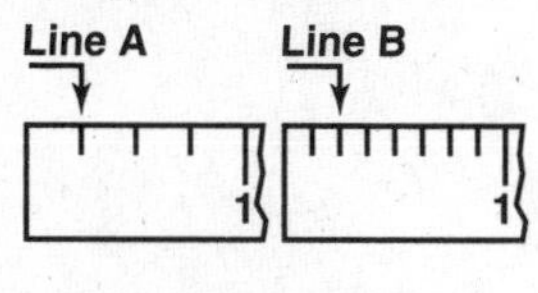


$\frac{1}{4} = \frac{2}{8}$

**5.** Line *C* and Line *D*

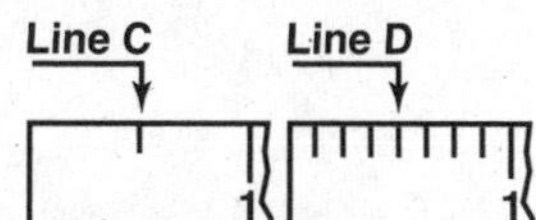


__________

**6.** Line *E* and Line *F*

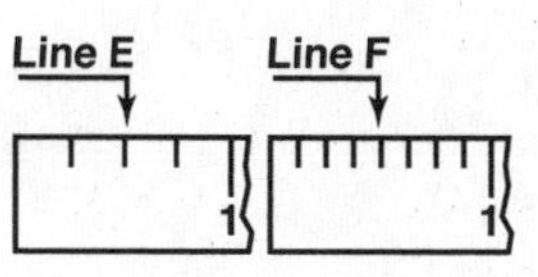


__________

**Check your answers on page 177.**

# Reducing Fractions

Fractions are reduced to lowest terms to make them easier to read and understand. **Reducing** a fraction to lowest terms means dividing both the numerator and denominator by the same number. A fraction is reduced to lowest terms when you can only divide both the numerator and denominator by 1.

**Reduce $\frac{12}{30}$ to lowest terms.**

**1.** Find a number that divides into 12 and 30 evenly.

$$\frac{12}{30} = \frac{12 \div 2}{30 \div 2} = \frac{6}{15}$$

**2.** Both 6 and 15 can still be divided by 3. The number is not in lowest terms. Reduce again.

$$\frac{6}{15} = \frac{6 \div 3}{15 \div 3} = \frac{2}{5}$$

**3.** See if you can divide both the numerator and the denominator by a number other than 1.

$\frac{2}{5}$ is in lowest terms.

**Reduce each fraction to lowest terms.**

**1.** $\frac{9}{21} =$

$$\frac{9}{21} = \frac{9 \div 3}{21 \div 3} = \frac{3}{7}$$

**2.** $\frac{8}{12} =$

**3.** $\frac{10}{18} =$

**4.** $\frac{10}{25} =$

**5.** $\frac{15}{20} =$

**6.** $\frac{7}{21} =$

**7.** $\frac{8}{10} =$

**8.** $\frac{20}{28} =$

**Write the number used to reduce each fraction to lowest terms.**

**9.** $\frac{9}{12} = \frac{9 \div \square}{12 \div \square} = \frac{3}{4}$

**10.** $\frac{7}{14} = \frac{7 \div \square}{14 \div \square} = \frac{1}{2}$

**11.** $\frac{6}{15} = \frac{6 \div \square}{15 \div \square} = \frac{2}{5}$

**12.** $\frac{12}{18} = \frac{12 \div \square}{18 \div \square} = \frac{2}{3}$

**13.** $\frac{20}{24} = \frac{20 \div \square}{24 \div \square} = \frac{5}{6}$

**14.** $\frac{9}{27} = \frac{9 \div \square}{24 \div \square} = 1$

**Check your answers on page 177.**

# Raising Fractions

You can multiply to change a fraction to an equivalent fraction with a larger denominator. This is called raising a fraction to higher terms.

**Raise $\frac{3}{4}$ to higher terms by multiplying the numerator and denominator by 5.**

1. Multiply the numerator by 5.

$$\frac{3}{4} = \frac{3 \times 5}{} = \frac{15}{}$$

2. Multiply the denominator by 5.

$$\frac{3}{4} = \frac{3 \times 5}{4 \times 5} = \frac{15}{20}$$

**Multiply the numerator and denominator by 2 to raise each fraction to higher terms.**

1. $\frac{2}{3} = \frac{2 \times 2}{3 \times 2} = \frac{4}{6}$
2. $\frac{3}{5} =$
3. $\frac{4}{7} =$
4. $\frac{5}{8} =$

**Write the number used to raise each fraction to higher terms.**

5. $\frac{6}{7} = \frac{12}{14}$ 2

   $\frac{6 \times 2}{7 \times 2} = \frac{12}{14}$
6. $\frac{4}{5} = \frac{16}{20}$ ___
7. $\frac{2}{9} = \frac{6}{27}$ ___
8. $\frac{2}{3} = \frac{10}{15}$ ___

**Raise each fraction to higher terms using the given numerator or denominator.**

9. $\frac{1}{4} = \frac{\boxed{6}}{24}$

   $\frac{1}{4} = \frac{1 \times 6}{4 \times 6} = \frac{6}{24}$
10. $\frac{3}{8} = \frac{\square}{16}$
11. $\frac{1}{2} = \frac{\square}{20}$
12. $\frac{7}{9} = \frac{\square}{27}$
13. $\frac{3}{4} = \frac{6}{\square}$
14. $\frac{5}{9} = \frac{15}{\square}$
15. $\frac{2}{3} = \frac{10}{\square}$
16. $\frac{1}{10} = \frac{10}{\square}$

**Check your answers on page 178.**

# Common Denominators

To compare some fractions, you may need to find a common denominator. A **common denominator** is a number that both denominators divide into evenly.

**Compare $\frac{2}{3}$ and $\frac{4}{5}$**

**1.** Multiply the two denominators to find a common denominator. $3 \times 5 = 15$. A common denominator for both fractions is 15.

$$\frac{2}{3} = \frac{\quad}{3 \times 5} = \frac{\quad}{15}$$

$$\frac{4}{5} = \frac{\quad}{5 \times 3} = \frac{\quad}{15}$$

**2.** Write each fraction in higher terms with 15 as the denominator.

$$\frac{2}{3} = \frac{2 \times 5}{3 \times 5} = \frac{10}{15}$$

$$\frac{4}{5} = \frac{4 \times 3}{5 \times 3} = \frac{12}{15}$$

**3.** Compare the numerators of the two new fractions. 12 is greater than 10.

$$\frac{12}{15} > \frac{10}{15} \text{ so } \frac{4}{5} > \frac{2}{3}$$

**Compare each set of fractions. Write >, <, or =.**

**1.** $\frac{2}{7}$ [>] $\frac{1}{4}$

$$\frac{2}{7} = \frac{2 \times 4}{7 \times 4} = \frac{8}{28}$$

$$\frac{1}{4} = \frac{1 \times 7}{4 \times 7} = \frac{7}{28}$$

**2.** $\frac{4}{7}$ ☐ $\frac{2}{3}$

**3.** $\frac{8}{16}$ ☐ $\frac{4}{8}$

**4.** $\frac{2}{3}$ ☐ $\frac{3}{5}$

**5.** $\frac{4}{9}$ ☐ $\frac{1}{2}$

**6.** $\frac{2}{5}$ ☐ $\frac{3}{8}$

**7.** $\frac{3}{4}$ ☐ $\frac{6}{9}$

**8.** $\frac{4}{5}$ ☐ $\frac{5}{6}$

**Answer the word problems.**

**9.** During summer vacation, Diana grew $\frac{1}{2}$ inch and Enrique grew $\frac{2}{3}$ inch. Who grew more?

**10.** A recipe uses $\frac{3}{8}$ cup of chopped onion and $\frac{1}{3}$ cup of chopped celery. Does the recipe use more onion or more celery?

**Check your answers on page 178.**

# Finding the Lowest Common Denominator

Another way to find a common denominator is by making a list of multiples of each denominator. A multiple of a number is the number multiplied by 1; 2, 3, and so on. For example, some multiples of 2 are 2, 4, 6, 8, and 10.

List the multiples for each denominator. Find the smallest number that appears on both lists of multiples. This number is the **lowest common denominator** (LCD) for the fractions.

**Compare $\frac{5}{6}$ and $\frac{3}{10}$.**

**1.** List the multiples of each denominator. The smallest number on both lists is 30. 30 is the LCD.

$\frac{5}{6}$ 6 12 18 24 [30] 36 42 48

$\frac{3}{10}$ 10 20 [30] 40 50

**2.** Raise each fraction to higher terms with 30 as the LCD.

$$\frac{5}{6} = \frac{5 \times 5}{6 \times 5} = \frac{25}{30}$$

$$\frac{3}{10} = \frac{3 \times 3}{10 \times 3} = \frac{9}{30}$$

**3.** Compare the numerators of the new fractions. 25 is greater than 9.

$$\frac{25}{30} > \frac{9}{30} \text{ so } \frac{5}{6} > \frac{3}{10}$$

**List the multiples of each denominator to find the LCD. Change the fractions. Then compare.**

**1.** $\frac{1}{12}$ [<] $\frac{11}{18}$

$\frac{1}{12}$ 12 24 [36] 48 60

$\frac{11}{18}$ 18 [36] 54

$$\frac{1}{12} = \frac{1 \times 3}{12 \times 3} = \frac{3}{36}$$

$$\frac{11}{18} = \frac{11 \times 2}{18 \times 2} = \frac{22}{36}$$

**2.** $\frac{8}{15}$ ☐ $\frac{2}{5}$

**3.** $\frac{2}{3}$ ☐ $\frac{4}{9}$

**4.** $\frac{2}{5}$ ☐ $\frac{7}{12}$

**5.** $\frac{7}{10}$ ☐ $\frac{3}{4}$

**6.** $\frac{5}{7}$ ☐ $\frac{3}{14}$

**7.** $\frac{1}{5}$ ☐ $\frac{1}{6}$

**8.** $\frac{7}{24}$ ☐ $\frac{5}{16}$

**Check your answers on page 178.**

# Changing Improper Fractions to Mixed Numbers

A **proper fraction** is a fraction with a numerator that is less than the denominator ($\frac{3}{4}$). An **improper fraction** is a fraction with a numerator that is greater than or equal to the denominator ($\frac{4}{4}$, $\frac{17}{3}$).

A **mixed number** has a whole number and a fraction part ($4\frac{1}{3}$). You may need to change an improper fraction to a whole or mixed number.

**Change $\frac{9}{4}$ to a mixed number.**

**1.** Set up a division problem. Divide the numerator, 9, by the denominator, 4.

$4\overline{)9}$

**2.** Divide.

$$\begin{array}{r} 2 \text{ R1} \\ 4\overline{)9} \\ -8 \\ \hline 1 \end{array}$$

**3.** Write the remainder, 1, over the divisor, 4.

$2\frac{1}{4}$

**Write *P* (proper fraction), *I* (improper fraction), *W* (whole number), or *M* (mixed number) on the line next to each problem.**

**1.** $4\frac{1}{5}$ = M **2.** $\frac{1}{8}$ = ___ **3.** $\frac{14}{3}$ = ___ **4.** 2 = ___ **5.** $\frac{7}{7}$ = ___

**Change each improper fraction to a whole number or mixed number.**

**6.** $\frac{11}{4} = 2\frac{3}{4}$

$$\begin{array}{r} 2 \text{ R3} = 2\frac{3}{4} \\ 4\overline{)11} \\ -8 \\ \hline 3 \end{array}$$

**7.** $\frac{30}{10}$ = **8.** $\frac{8}{3}$ = **9.** $\frac{21}{5}$ = **10.** $\frac{16}{7}$ =

**11.** $\frac{15}{8}$ = **12.** $\frac{9}{2}$ = **13.** $\frac{13}{6}$ = **14.** $\frac{20}{4}$ = **15.** $\frac{25}{8}$ =

**Check your answers on page 178.**

# Changing Mixed Numbers to Improper Fractions

You have changed an improper fraction to a whole or mixed number. You can also change a mixed number or whole number to an improper fraction.

**Change $3\frac{2}{7}$ to an improper fraction.**

1. Write a fraction with the same denominator, 7.

$$3\frac{2}{7} = \frac{\phantom{0}}{7}$$

2. Multiply the denominator, 7, by the whole number, 3.

$$7 \times 3 = 21$$

3. Add the numerator, 2, to 21. Write the sum over the denominator, 7.

$$21 + 2$$

$$\frac{23}{7}$$

**Change each mixed or whole number to an improper fraction.**

1. $5\frac{1}{2} = \frac{11}{2}$
   $5 \times 2 = 10$
   $10 + 1 = 11$
2. $8\frac{2}{3} =$
3. $6 =$
4. $12\frac{3}{4} =$
5. $1\frac{5}{8} =$
6. $3\frac{3}{5} =$
7. $7\frac{1}{2} =$
8. $9\frac{1}{7} =$
9. $10\frac{2}{3} =$
10. $5\frac{5}{6} =$
11. $20 =$
12. $11\frac{1}{3} =$
13. $10\frac{2}{5} =$
14. $33\frac{1}{2} =$
15. $100 =$

**Answer the word problems.**

16. A recipe calls for $2\frac{1}{4}$ cups of flour. Ali has only a $\frac{1}{4}$-cup measuring cup. She writes $2\frac{1}{4}$ as an improper fraction. What improper fraction does Ali write?

17. Bart worked for $5\frac{3}{4}$ hours yesterday. How is $5\frac{3}{4}$ written as an improper fraction?

**Check your answers on page 179.**

# Adding Fractions with Like Denominators

To add fractions with the same denominator, add only the numerators. Keep the same denominator. Always reduce the answer to lowest terms.

**Add:** $\frac{2}{8} + \frac{4}{8}$

1. The denominators are the same. Write the denominator under the fraction bar.

$\frac{2}{8} + \frac{4}{8} = \frac{}{8}$

2. Add only the numerators. $2 + 4 = 6$. Write the sum over the denominator.

$\frac{2}{8} + \frac{4}{8} = \frac{6}{8}$

3. If the answer is not in lowest terms, reduce it.

$\frac{6}{8} = \frac{6 \div 2}{8 \div 2} = \frac{3}{4}$

**Add. If the answer is an improper fraction, change it to a mixed number. Reduce the answer if possible.**

1. $\frac{4}{5} + \frac{3}{5} = \frac{7}{5} = 1\frac{2}{5}$
2. $\frac{6}{10} + \frac{5}{10} =$
3. $\frac{2}{9} + \frac{4}{9} =$
4. $\frac{9}{5} + \frac{4}{5} =$
5. $\frac{7}{3} + \frac{4}{3} =$
6. $\frac{3}{6} + \frac{2}{6} =$
7. $\frac{4}{12} + \frac{7}{12} =$
8. $\frac{8}{24} + \frac{12}{24} =$
9. $\frac{6}{4} + \frac{8}{4} =$

**Answer the word problems.**

10. It is $\frac{2}{10}$ mile from the stop sign to the gas station. It is $\frac{6}{10}$ mile from the gas station to the traffic signal. How far is it from the stop sign to the traffic signal?

11. Sharon mixed $\frac{1}{3}$ cup of pineapple juice with $\frac{2}{3}$ cup of orange juice. After she mixed the juices, how much juice did Sharon have?

**Check your answers on page 179.**

# Adding Mixed Numbers with Like Denominators

Mixed numbers with the same denominator are added as you would add fractions. First, add the fractions. Then add any whole numbers.

**Add:** $3\frac{3}{8} + 1\frac{7}{8}$

1. The denominators are the same. Write the denominator. Add the numerators.

$$\begin{array}{r} 3\frac{3}{8} \\ +1\frac{7}{8} \\ \hline \frac{10}{8} \end{array}$$

2. Add the whole numbers.

$$\begin{array}{r} 3\frac{3}{8} \\ +1\frac{7}{8} \\ \hline 4\frac{10}{8} \end{array}$$

3. Change any improper fractions to mixed numbers. Add and reduce.

$$4\frac{10}{8} = 4 + 1\frac{2}{8} =$$

$$5\frac{2}{8} = 5\frac{1}{4}$$

**Add. Change improper fractions to mixed numbers. Reduce the answer if possible.**

1. $6\frac{6}{10} + 3\frac{5}{10}$

   $9\frac{11}{10} = 9 + 1\frac{1}{10}$

   $= 10\frac{1}{10}$

2. $3\frac{2}{9} + 2\frac{1}{9}$

3. $9\frac{9}{10} + 15\frac{5}{10}$

4. $3\frac{3}{5} + 4\frac{4}{5}$

5. $2\frac{7}{12} + 7\frac{5}{12}$

6. $3\frac{3}{4} + 1\frac{3}{4}$

7. $4\frac{1}{8} + 5\frac{3}{8}$

8. $8\frac{5}{6} + 2\frac{1}{6}$

9. $5\frac{1}{2} + 7$

10. $12\frac{7}{8} + 3\frac{3}{8}$

**Answer the word problems.**

11. Brad bought $2\frac{1}{4}$ pounds of carrots and $1\frac{1}{4}$ pounds of green beans. How many pounds of vegetables did he buy in all?

12. Lenore needs $3\frac{1}{8}$ yards of material for a jacket, 4 yards for a dress, and $\frac{5}{8}$ yard for a scarf. How many yards of material does she need in all?

**Check your answers on page 179.**

# Adding Fractions with Unlike Denominators

To add fractions with unlike denominators, find the LCD of the fractions. See page 117 for help.

**Add:** $\frac{3}{4} + \frac{2}{3}$

1. Find the LCD. The lowest common denominator for both fractions is 12.

$$\frac{3}{4} = \frac{\quad}{12} \qquad \frac{2}{3} = \frac{\quad}{12}$$

2. Raise each fraction to higher terms with 12 as the denominator. Add the fractions.

$$\frac{3}{4} = \frac{9}{12} \qquad \frac{2}{3} = \frac{8}{12}$$

$$\frac{9}{12} + \frac{8}{12} = \frac{17}{12}$$

3. Change the improper fraction to a mixed number. Reduce the answer if possible.

$$\frac{17}{12} = 1\frac{5}{12}$$

**Add. Reduce the answer if possible.**

1. $\frac{2}{9} = \frac{4}{18}$
   $+\frac{5}{18} = \frac{5}{18}$
   $\frac{9}{18} = \frac{9 \div 9}{18 \div 9} = \frac{1}{2}$

2. $\frac{1}{3} + \frac{6}{12}$

3. $\frac{3}{8} + \frac{15}{16}$

4. $\frac{1}{4} + \frac{2}{7}$

5. $\frac{1}{2} + \frac{7}{8} =$

6. $\frac{2}{5} + \frac{3}{10} =$

7. $\frac{5}{8} + \frac{7}{12} =$

8. $\frac{5}{6} + \frac{3}{12} =$

**Answer the word problems.**

9. The plywood on a floor is $\frac{5}{8}$ inch thick. The tile on top of the plywood is $\frac{5}{16}$ inch thick. What is the thickness of the plywood and tile together?

10. In a company, $\frac{1}{6}$ of the workers are under 25 years old, and $\frac{1}{4}$ of the workers are between the ages of 25 and 40. What fraction of the workers are under 40 years old?

**Check your answers on page 179.**

# Adding Mixed Numbers with Unlike Denominators

Mixed numbers with different denominators can also be added using the LCD.

**Add:** $2\frac{3}{4} + 1\frac{1}{3}$

1. Find the LCD. A common denominator for both fractions is 12.

$$\begin{aligned} 2\tfrac{3}{4} &= 2\tfrac{}{12} \\ +1\tfrac{1}{3} &= 1\tfrac{}{12} \end{aligned}$$

2. Raise each fraction to higher terms with 12 as the denominator. Add the fractions. Add the whole numbers.

$$\begin{aligned} 2\tfrac{3}{4} &= 2\tfrac{9}{12} \\ +1\tfrac{1}{3} &= 1\tfrac{4}{12} \\ & \quad 3\tfrac{13}{12} \end{aligned}$$

3. Change the improper fraction in the answer to a mixed number. Add the new mixed number to the whole number. Reduce the answer if possible.

$$3\tfrac{13}{12} = 3 + 1\tfrac{1}{12} = 4\tfrac{1}{12}$$

**Add. Reduce the answer if possible.**

1. $4\frac{4}{5} = 4\frac{16}{20}$
   $+3\frac{3}{4} = 3\frac{15}{20}$
   $7\frac{31}{20} = 8\frac{11}{20}$

2. $9\frac{2}{3} + 7\frac{1}{5}$

3. $1\frac{5}{8} + 8\frac{3}{10}$

4. $5\frac{7}{9} + 5\frac{1}{4}$

5. $2\frac{9}{16} + 4\frac{2}{3}$

6. $6\frac{5}{6} + 3\frac{3}{4}$

7. $2\frac{1}{3} + 10\frac{5}{12}$

8. $4\frac{3}{5} + 8\frac{7}{10}$

**Answer the word problems.**

9. Rory jogged $2\frac{1}{2}$ miles on Saturday morning and $1\frac{3}{10}$ miles Sunday evening. How many miles did Rory jog in all?

10. Pat mailed a package that weighed $6\frac{1}{4}$ pounds and another that weighed $3\frac{1}{2}$ pounds. How much did the packages weigh all together?

**Check your answers on page 180.**

# Subtracting Fractions with Like Denominators

Subtracting fractions with like denominators is similar to adding fractions with like denominators. Subtract only the numerators.

**Subtract:** $\frac{7}{10} - \frac{1}{10}$

**1.** The denominators are the same. Write the denominator under the fraction bar.

$$\frac{7}{10} - \frac{1}{10} = \frac{\phantom{0}}{10}$$

**2.** Subtract only the numerators. $7 - 1 = 6$. Write the difference over the denominator.

$$\frac{7}{10} - \frac{1}{10} = \frac{6}{10}$$

**3.** If the answer is not in lowest terms, reduce it.

$$\frac{6}{10} = \frac{6 \div 2}{10 \div 2} = \frac{3}{5}$$

**Subtract. Reduce answers if possible.**

1. $\frac{3}{4} - \frac{1}{4}$; $\frac{2}{4} = \frac{2 \div 2}{4 \div 2} = \frac{1}{2}$
2. $\frac{5}{6} - \frac{1}{6}$
3. $\frac{7}{8} - \frac{3}{8}$
4. $\frac{9}{10} - \frac{3}{10}$
5. $\frac{2}{3} - \frac{1}{3}$

6. $\frac{6}{7} - \frac{3}{7} =$
7. $\frac{11}{18} - \frac{7}{18} =$
8. $\frac{4}{5} - \frac{2}{5} =$
9. $\frac{7}{12} - \frac{1}{12} =$

**Answer the word problems.**

10. Tana's gas tank is $\frac{3}{4}$ full. She uses $\frac{1}{4}$ tank of gas to go to the airport and back home. When Tana gets home, what fraction of the gas tank is full?

11. Carl lives $\frac{8}{10}$ of a mile from the library. He lives $\frac{4}{10}$ of a mile from the school. How much closer does Carl live to the school than to the library?

**Check your answers on page 180.**

# Subtracting Mixed Numbers with Like Denominators

Subtract mixed numbers with the same denominator the same way you subtract fractions. First, subtract the fractions. Then subtract the whole numbers.

**Subtract:** $\begin{array}{r} 8\frac{7}{12} \\ -6\frac{5}{12} \\ \hline \end{array}$

**1.** The denominators are the same. Write the denominator in the answer. Subtract the numerators.

$$\begin{array}{r} 8\frac{7}{12} \\ -6\frac{5}{12} \\ \hline \frac{2}{12} \end{array}$$

**2.** Subtract the whole numbers.

$$\begin{array}{r} 8\frac{7}{12} \\ -6\frac{5}{12} \\ \hline 2\frac{2}{12} \end{array}$$

**3.** Reduce the answer to lowest terms if possible.

$$2\frac{2}{12} = 2\frac{1}{6}$$

**Subtract. Reduce answers if possible.**

1. $\begin{array}{r} 6\frac{7}{20} \\ -5\frac{3}{20} \\ \hline \end{array}$ $1\frac{4}{20} = 1\frac{1}{5}$

2. $\begin{array}{r} 3\frac{6}{7} \\ -1\frac{4}{7} \\ \hline \end{array}$

3. $\begin{array}{r} 12\frac{11}{15} \\ -7\frac{2}{15} \\ \hline \end{array}$

4. $\begin{array}{r} 8\frac{7}{16} \\ -8\frac{1}{16} \\ \hline \end{array}$

5. $\begin{array}{r} 9\frac{5}{6} \\ -\frac{1}{6} \\ \hline \end{array}$

6. $\begin{array}{r} 15\frac{7}{10} \\ -13\frac{3}{10} \\ \hline \end{array}$

7. $\begin{array}{r} 6\frac{3}{5} \\ -1\frac{2}{5} \\ \hline \end{array}$

8. $\begin{array}{r} 5\frac{7}{9} \\ -\frac{4}{9} \\ \hline \end{array}$

9. $\begin{array}{r} 8\frac{7}{8} \\ -3\frac{3}{8} \\ \hline \end{array}$

10. $\begin{array}{r} 13\frac{2}{3} \\ -13\frac{1}{3} \\ \hline \end{array}$

**Answer the word problems.**

11. A canister contains $3\frac{3}{4}$ cups of flour. How much flour will be left in the canister if $2\frac{1}{4}$ cups are used for baking?

12. A board measures $25\frac{7}{8}$ inches in length. If an $8\frac{5}{8}$-inch piece is cut from the board, how long will the board be?

**Check your answers on page 180.**

# Subtracting Fractions with Unlike Denominators

The LCD can be used to subtract fractions with different denominators. Always reduce the answer to lowest terms.

**Subtract:** $\frac{3}{5} - \frac{1}{4}$

**1.** Find a common denominator. A common denominator for both fractions is 20.

$$\begin{array}{r} \frac{3}{5} = \frac{\phantom{00}}{20} \\ -\frac{1}{4} = \frac{\phantom{00}}{20} \\ \hline \end{array}$$

**2.** Raise each fraction to higher terms with 20 as the denominator.

$$\begin{array}{r} \frac{3}{5} = \frac{12}{20} \\ -\frac{1}{4} = \frac{5}{20} \\ \hline \end{array}$$

**3.** Subtract. The answer is in lowest terms.

$$\begin{array}{r} \frac{3}{5} = \frac{12}{20} \\ -\frac{1}{4} = \frac{5}{20} \\ \hline \frac{7}{20} \end{array}$$

**Subtract. Reduce answers if possible.**

1. $\begin{array}{r} \frac{1}{3} = \frac{4}{12} \\ -\frac{1}{4} = \frac{3}{12} \\ \hline \frac{1}{12} \end{array}$

2. $\begin{array}{r} \frac{6}{6} \\ -\frac{1}{5} \\ \hline \end{array}$

3. $\begin{array}{r} \frac{4}{7} \\ -\frac{1}{2} \\ \hline \end{array}$

4. $\begin{array}{r} \frac{5}{8} \\ -\frac{1}{3} \\ \hline \end{array}$

5. $\begin{array}{r} \frac{11}{12} \\ -\frac{5}{9} \\ \hline \end{array}$

6. $\frac{7}{8} - \frac{1}{3} =$

7. $\frac{9}{10} - \frac{3}{5} =$

8. $\frac{11}{16} - \frac{1}{4} =$

9. $\frac{3}{4} - \frac{2}{5} =$

10. $\frac{6}{7} - \frac{2}{3} =$

**Answer the word problems.**

11. Chee bought a can of mixed nuts weighing $\frac{1}{2}$ pound. He used $\frac{1}{8}$ pound to bake cookies. How much of the mixed nuts was left over?

12. Yuko spends about $\frac{1}{3}$ of each weekday sleeping. She also spends about $\frac{1}{12}$ of each weekday doing homework. How much more of the day does she spend sleeping than doing homework?

**Check your answers on page 181.**

## Subtracting Fractions and Mixed Numbers with Unlike Denominators

Subtract mixed numbers with unlike numbers using the LCD.

**Subtract:** $6\frac{3}{4} - 3\frac{1}{3}$

1. Find the LCD. The LCD for both fractions is 12.

$$\begin{array}{r} 6\frac{3}{4} = \frac{\phantom{0}}{12} \\ -\,3\frac{1}{3} = \frac{\phantom{0}}{12} \\ \hline \end{array}$$

2. Raise each fraction to higher terms with 12 as the denominator. Subtract the fractions.

$$\begin{array}{r} 6\frac{3}{4} = 6\frac{9}{12} \\ -\,3\frac{1}{3} = 3\frac{4}{12} \\ \hline \frac{5}{12} \end{array}$$

3. Subtract the whole numbers. Reduce the answer if possible.

$$\begin{array}{r} 6\frac{3}{4} = 6\frac{9}{12} \\ -\,3\frac{1}{3} = 3\frac{4}{12} \\ \hline 3\frac{5}{12} \end{array}$$

**Subtract. Reduce answers if possible.**

1. $$\begin{array}{r} 8\frac{5}{6} = 8\frac{30}{36} \\ -\,4\frac{1}{9} = 4\frac{4}{36} \\ \hline 4\frac{26}{36} = 4\frac{13}{18} \end{array}$$

2. $$\begin{array}{r} 11\frac{3}{5} \\ -\,9\frac{1}{3} \\ \hline \end{array}$$

3. $$\begin{array}{r} 7\frac{4}{5} \\ -\,4\frac{1}{6} \\ \hline \end{array}$$

4. $$\begin{array}{r} 2\frac{5}{8} \\ -\,\frac{2}{5} \\ \hline \end{array}$$

5. $$\begin{array}{r} 15\frac{13}{16} \\ -\,13\frac{3}{5} \\ \hline \end{array}$$

6. $$\begin{array}{r} 4\frac{4}{5} = 4\frac{32}{40} \\ -\,2\frac{5}{8} = 2\frac{25}{40} \\ \hline 2\frac{7}{40} \end{array}$$

7. $$\begin{array}{r} 8\frac{7}{9} \\ -\,3\frac{1}{3} \\ \hline \end{array}$$

8. $$\begin{array}{r} 12\frac{9}{10} \\ -\,10\frac{7}{8} \\ \hline \end{array}$$

9. $$\begin{array}{r} 5\frac{9}{16} \\ -\,1\frac{1}{2} \\ \hline \end{array}$$

10. $$\begin{array}{r} 7\frac{5}{12} \\ -\,6\frac{1}{10} \\ \hline \end{array}$$

**Answer the word problems.**

11. Matt has a piece of string $65\frac{7}{8}$ inches long. He cuts off a piece of string $10\frac{1}{4}$ inches for his lacrosse stick. After Matt cuts the string, how much is left?

12. A roll of landscaping plastic is $32\frac{5}{8}$ yards long. Rachel used $12\frac{1}{4}$ yards in her garden. How much plastic was left on the roll?

**Check your answers on page 181.**

## Reading Line Graphs

A line graph connects points that show changes over time. A line graph has two axes. The horizontal axis goes side to side and measures time. The vertical axis goes up and down and measures the amount of something.

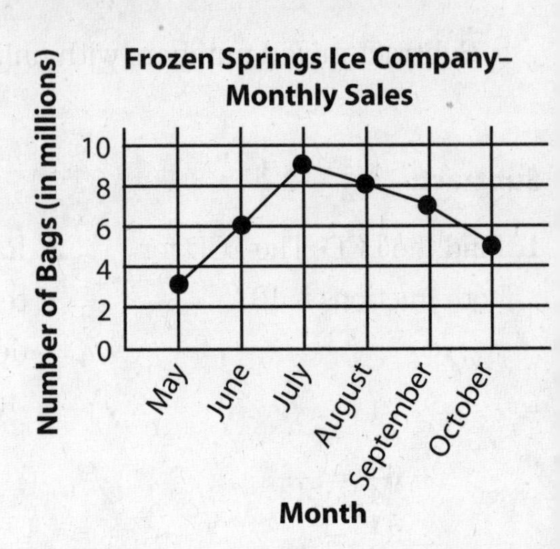

**Strategy** Learn how to read points on a line graph. Follow these steps.

How many bags of ice were sold in July?

**1.** Study the title and labels of the line graph.

**2.** Find the measure of time on the horizontal axis, and find the point directly above it.

Find *July* on the horizontal axis. Find the point directly above the label for July.

**3.** From that point, read across the line graph to find the amount on the vertical axis.

Read straight across to the vertical axis. The point is halfway between 8 and 10 on the vertical axis. The point for July is at 9.

Since the amounts on the vertical scale are in millions, 9 million bags of ice were sold in July.

**Exercise 1: Use the line graph to solve problems 1–3.**

**1.** In which year did the company have the greatest number of employees?

_______________________________

**2.** In which year did the company have the fewest employees?

_______________________________

**3.** How many more employees did the company have in year 4 than in year 3?

_______________________________

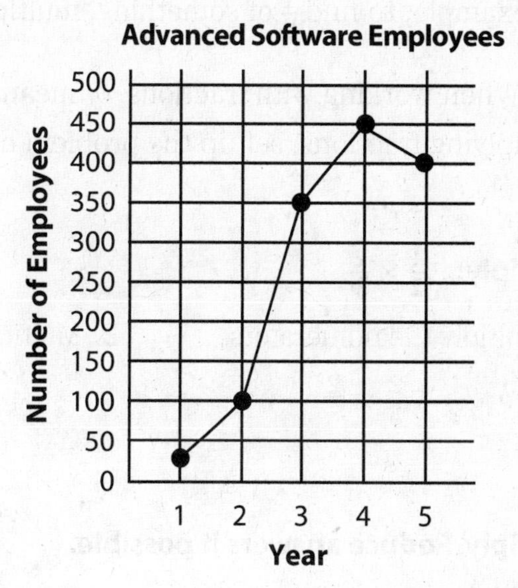


**Exercise 2: Use the line graph to solve problems 4–6.**

**4.** How many campsites are reserved during week 2?

_______________________________

**5.** During what week are the fewest campsites reserved?

_______________________________

**6.** How many reservations are there in all for weeks 3, 4, and 5?

_______________________________

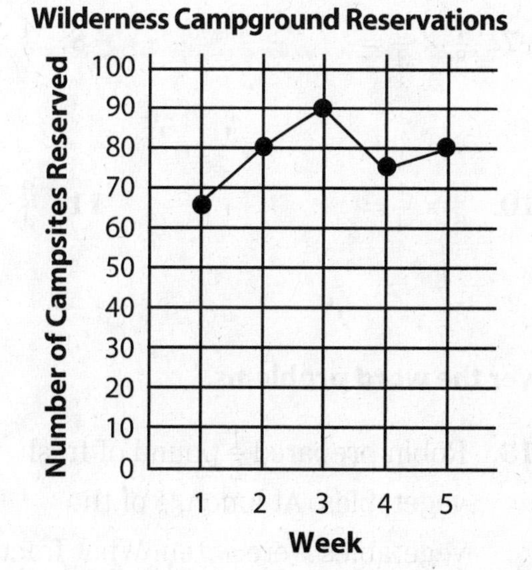


**Check your answers on page 182.**

# Multiplying Fractions by Fractions

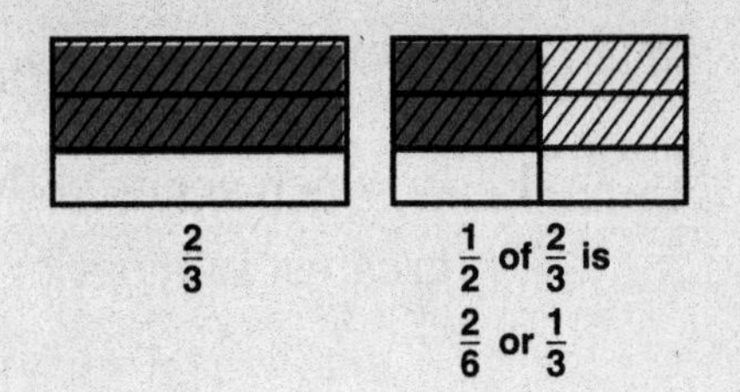


Multiplying fractions is one way to find a part of an amount. For example, to find $\frac{1}{2}$ of something, multiply by $\frac{1}{2}$.

When working with fractions, *of* means "to multiply". When multiplying fractions, set up the problem horizontally.

**Multiply:** $\frac{1}{2} \times \frac{2}{3}$

**1.** Multiply the numerators.

$\frac{1}{2} \times \frac{2}{3} = \frac{2}{}$

**2.** Multiply the denominators.

$\frac{1}{2} \times \frac{2}{3} = \frac{2}{6}$

**3.** Reduce to lowest terms.

$\frac{2}{6} = \frac{1}{3}$

**Multiply. Reduce answers if possible.**

**1.** $\frac{1}{3} \times \frac{3}{4} =$

$\frac{1 \times 3}{3 \times 4} = \frac{3}{12} = \frac{1}{4}$

**2.** $\frac{5}{8} \times \frac{1}{3} =$

**3.** $\frac{1}{2} \times \frac{3}{4} =$

**4.** $\frac{2}{5} \times \frac{7}{12} =$

**5** $\frac{3}{16} \times \frac{1}{5} =$

**6.** $\frac{7}{10} \times \frac{5}{6} =$

**7.** $\frac{1}{8} \times \frac{2}{3} =$

**8.** $\frac{4}{7} \times \frac{5}{12} =$

**9.** $\frac{3}{20} \times \frac{3}{5} =$

**10.** $\frac{5}{6} \times \frac{1}{4} =$

**11.** $\frac{3}{4} \times \frac{2}{9} =$

**12.** $\frac{1}{2} \times \frac{3}{8} =$

**Answer the word problems.**

**13.** Robin prepared $\frac{1}{2}$ pound of fresh vegetables. At lunch, $\frac{1}{3}$ of the vegetables were eaten. What fraction of the vegetables was eaten?

**14.** In Mr. Roma's class, $\frac{3}{8}$ of the students are men, and $\frac{1}{2}$ of those men are married. What fraction of the class is married men?

**Check your answers on page 182.**

# Multiplying Mixed Numbers and Whole Numbers

To multiply mixed and whole numbers, change both numbers to improper fractions. Then multiply as you do with fractions. Reduce to lowest terms if necessary.

**Multiply:** $4\frac{1}{2} \times 6$

**1.** Write $4\frac{1}{2}$ as an improper fraction.

$4\frac{1}{2} = \frac{4 \times 2 + 1}{2} = \frac{9}{2}$

**2.** Write 6 as an improper fraction.

$6 = \frac{6}{1}$

**3.** Multiply. Change the improper fraction to a whole number. The answer is in lowest terms.

$\frac{9}{2} \times \frac{6}{1} = \frac{54}{2} = 27$

**Multiply. Change improper fraction answers to mixed numbers. Reduce answers if possible.**

**1.** $3\frac{1}{4} \times 2 =$

$\frac{13}{4} \times \frac{2}{1} = \frac{13 \times 2}{4 \times 1} = \frac{26}{4}$

$\frac{26}{4} = 6\frac{2}{4} = 6\frac{1}{2}$

**2.** $2\frac{1}{2} \times 3 =$

**3.** $1\frac{2}{5} \times 4 =$

**4.** $3 \times 1\frac{2}{5} =$

**5.** $4\frac{1}{4} \times 3\frac{1}{2}$

**6.** $2\frac{1}{3} \times 1\frac{2}{5} =$

**7.** $3\frac{1}{2} \times 2\frac{2}{7} =$

**8.** $1\frac{5}{12} \times 7 =$

**9.** $7 \times 9\frac{1}{3} =$

**10.** $5\frac{1}{2} \times 8 =$

**11.** $1\frac{5}{8} \times 2\frac{2}{3} =$

**12.** $2\frac{3}{8} \times 2\frac{4}{5} =$

**Answer the word problems.**

**13.** Rita exercises regularly. Each morning, she swims $\frac{1}{2}$ mile. How many miles does Rita swim each week? (1 week = 7 days)

**14.** On Monday, 6 plants were in bloom. On Friday, $3\frac{1}{2}$ times as many plants were blooming. How many plants were blooming on Friday?

**Check your answers on page 182.**

# Dividing Fractions by Fractions

To find out how many equal parts are in a fraction, you will need to divide a fraction by a fraction.

**Divide:** $\frac{2}{3} \div \frac{1}{5}$

**1. Invert**, or turn upside down, the fraction to the right of the division sign.

$\frac{1}{5} \rightarrow \frac{5}{1}$

**2.** Multiply by the new fraction.

$\frac{2}{3} \times \frac{5}{1} = \frac{10}{3}$

**3.** If the answer is an improper fraction, change it to a mixed number.

$\frac{10}{3} = 3\frac{1}{3}$

**Divide. Change improper fraction answers to mixed numbers.**
**Reduce answers if possible.**

**1.** $\frac{1}{2} \div \frac{7}{8}$

$\frac{1}{2} \times \frac{8}{7} = \frac{1 \times 8}{2 \times 7} = \frac{8}{14}$

$\frac{8}{14} = \frac{4}{7}$

**2.** $\frac{2}{3} \div \frac{3}{4}$

**3.** $\frac{3}{8} \div \frac{1}{6}$

**4.** $\frac{7}{12} \div \frac{2}{3}$

**5.** $\frac{1}{3} \div \frac{5}{8}$

**6.** $\frac{5}{6} \div \frac{2}{5}$

**7.** $\frac{1}{7} \div \frac{1}{4}$

**8.** $\frac{8}{15} \div \frac{1}{2}$

**9.** $\frac{3}{5} \div \frac{13}{20}$

**10.** $\frac{3}{8} \div \frac{1}{4}$

**11.** $\frac{4}{5} \div \frac{9}{10}$

**12.** $\frac{2}{7} \div \frac{3}{13}$

**Answer the word problems.**

**13.** For a sewing project, Lara needs to divide $\frac{3}{4}$ yard of material into $\frac{1}{4}$-yard pieces. How many pieces will she get?

**14.** If Lara divides $\frac{3}{4}$ yard of material into $\frac{1}{8}$-yard pieces, how many pieces will she get?

**Check your answers on page 182.**

# Dividing Mixed Numbers and Whole Numbers

Some problems involve division with whole numbers and mixed numbers. To solve these problems, change whole or mixed numbers to improper fractions before dividing.

**Divide:** $7\frac{1}{2} \div 2$

**1.** Change both numbers to improper fractions.

$7\frac{1}{2} = \frac{7 \times 2 + 1}{2} = \frac{15}{2}$

$2 = \frac{2}{1}$

**2.** Invert the fraction to the right of the division sign.

$\frac{15}{2} \div \frac{2}{1} = \frac{15}{2} \times \frac{1}{2}$

**3.** Multiply the new fractions. Change the improper fraction to a mixed number.

$\frac{15}{2} \times \frac{1}{2} = \frac{15}{4} = 3\frac{3}{4}$

**Divide. Reduce the answer if possible.**

**1.** $4\frac{1}{2} \div 2\frac{1}{4} =$

$\frac{9}{2} \div \frac{9}{4} = \frac{9}{2} \times \frac{4}{9} = \frac{9 \times 4}{2 \times 9} = \frac{36}{18}$

$\frac{36}{18} = 2$

**2.** $2\frac{2}{3} \div 1\frac{1}{3} =$

**3.** $9\frac{3}{8} \div 3 =$

**4.** $12\frac{4}{5} \div 3\frac{1}{5} =$

**5.** $4 \div 3\frac{1}{2} =$

**6.** $10\frac{2}{5} \div 1\frac{1}{5} =$

**7.** $4 \div 2\frac{3}{4} =$

**8.** $8\frac{1}{3} \div 1\frac{1}{4} =$

**Answer the word problems.**

**9.** A produce clerk must divide $2\frac{1}{2}$ pounds of chives into packages that each contain $\frac{1}{8}$ pound. How many packages can be made?

**10.** A 25-pound block of cheese is to be cut into pieces that each weigh $\frac{1}{2}$ pound. How many pieces of cheese can be cut from the block?

**Check your answers on page 183.**

# Percents as Fractions

Use what you know about fractions and decimals to change percents to fractions and fractions to percents.

**Change 425% to a mixed number.**

Write the number over 100 without the percent sign. Change to a mixed number and reduce.

$$425\% = \frac{425}{100} = 4\frac{25}{100} = 4\frac{1}{4}$$

**Change $\frac{3}{4}$ to a percent.**

First change the fraction to a decimal. Divide the numerator by the denominator. Then move the decimal point two places to the right. Add a percent sign.

$$\frac{3}{4} = 3 \div 4 = 0.75 = 75\%$$

**Change each percent to a fraction, mixed number, or whole number. Reduce answers if possible.**

1. $35\% =$ $\frac{35}{100} = \frac{7}{20}$
2. $500\% =$
3. $4\% =$
4. $55\% =$
5. $250\% =$
6. $70\% =$
7. $30\% =$
8. $260\% =$

**Change each fraction to a percent.**

9. $\frac{2}{3} =$

$$\begin{array}{r} .66\frac{2}{3} \\ 3\overline{)2.00} \\ \underline{-1\ 8} \\ 20 \\ \underline{-18} \\ 2 \end{array}$$

$.66\frac{2}{3} = 66\frac{2}{3}\%$

10. $\frac{4}{5} =$
11. $\frac{9}{20} =$
12. $\frac{5}{6} =$

**Answer the word problems.**

13. A small grocery store lost $\frac{1}{3}$ of its customers when a large supermarket opened nearby. Write the fraction of customers lost as a percent.

14. A building supply store increased its sales by 130% for the year by offering special sale prices. Write the percent of increased sales as a mixed number.

**Check your answers on page 183.**

# Ratios

A **ratio** is a fraction that shows a relationship between two numbers. For example, if there are 6 men and 15 women, then the ratio of men to women can be shown as the fraction $\frac{6}{15}$.

A ratio may be reduced without changing the meaning. It can have a denominator of 1, but it cannot be changed to a whole number or a mixed number.

**Write a ratio to show that a car traveled 60 miles in 2 hours.**

1. Write the first number in the relationship as the numerator of the fraction.

   $\frac{60}{\phantom{2}}$

2. Write the second number as the denominator.

   $\frac{60}{2}$

3. Reduce.

   $\frac{60 \div 2}{2 \div 2} = \frac{30}{1}$

**Write each ratio as a fraction.**

1. 2 out of 5 voters

   $\frac{2}{5}$

2. 9 out of 10 adults

3. 4 losses in 7 games

4. 3 dollars per pound

   $\frac{3}{1}$

5. 56 words per minute

6. 28 miles per gallon

**Reduce each ratio to lowest terms.**

7. $\frac{12}{16}$

   $\frac{12}{16} = \frac{12 \div 4}{16 \div 4} = \frac{3}{4}$

8. $\frac{3}{9}$

9. $\frac{10}{4}$

10. $\frac{16}{20}$

11. $\frac{25}{5}$

**Write a ratio for each word problem. Reduce if possible.**

12. Milli and James invited 72 people to their wedding. Only 64 people attended. Write the ratio of the number of people who attended to the number of people who were invited.

13. Laura delivers 42 newspapers each weekday and 56 newspapers on Sunday. What is the ratio of the number of newspapers she delivers on a Sunday to the number of newspapers she delivers on a weekday?

**Check your answers on page 184.**

# Equal Ratios

**Equal ratios** are equal fractions. You can change ratios to higher terms by multiplying both the numerator and the denominator by the same number. You can reduce ratios to lower terms by dividing both the numerator and the denominator by the same number.

You can cross-multiply to check if two ratios are equal. If the ratios are equal, the answers to the cross-multiplication will be equal.

**Change $\frac{3}{4}$ to an equal ratio with 8 as the denominator.** $\frac{3}{4} = \frac{\square}{8}$

1. Look at the denominators. The 4 is multiplied by 2 to get the 8.

$$\frac{3}{4} = \frac{}{4 \times 2} = \frac{\square}{8}$$

2. Multiply the numerator by the same number, 2.

$$\frac{3}{4} = \frac{3 \times 2}{4 \times 2} = \frac{6}{8}$$

3. Cross multiply to check.

$$\frac{3}{4} \times \frac{6}{8}$$

$$3 \times 8 = 4 \times 6$$

$$24 = 24$$

**Complete each pair of equal ratios.**

1. $\frac{1}{5} = \frac{\boxed{2}}{10}$

   $\frac{1}{5} = \frac{1 \times 2}{5 \times 2} = \frac{2}{10}$

   Check: $\frac{1}{5} \times \frac{2}{10}$

   $1 \times 10 = 5 \times 2$

   $10 = 10$

2. $\frac{5}{12} = \frac{\square}{24}$

3. $\frac{1}{4} = \frac{\square}{12}$

4. $\frac{4}{7} = \frac{\square}{35}$

5. $\frac{5}{8} = \frac{20}{\boxed{32}}$

   $\frac{5}{8} = \frac{5 \times 4}{8 \times 4} = \frac{20}{32}$

6. $\frac{4}{15} = \frac{8}{\square}$

7. $\frac{1}{3} = \frac{6}{\square}$

8. $\frac{7}{8} = \frac{21}{\square}$

**Cross-multiply to check if the ratios are equal.**

9. $\frac{6}{8}$ and $\frac{12}{20}$

   $6 \times 20 = 120$

   $8 \times 12 = 96$

   120 and 96

   not equal

10. $\frac{4}{6}$ and $\frac{20}{30}$

11. $\frac{12}{15}$ and $\frac{4}{5}$

12. $\frac{4}{12}$ and $\frac{1}{4}$

**Check your answers on page 184.**

# Solving Proportions

Two equal ratios are called a **proportion**. You can cross multiply to solve a proportion for the unknown number.

**Solve for *n*.** $\frac{3}{6} = \frac{n}{10}$

1. Cross multiply. $6n$ means 6 times $n$.

$$\frac{3}{8} \times \frac{n}{10}$$
$$6 \times n = 3 \times 10$$
$$6n = 30$$

2. Divide both sides of the equation by the number next to $n$, 6.

$$6n = 30$$
$$\frac{6n}{6} = \frac{30}{6}$$
$$n = 5$$

3. Check by substituting the answer, 5, for $n$. Cross multiply.

$$\frac{3}{6} = \frac{5}{10}$$
$$6 \times 5 = 3 \times 10$$
$$30 = 30$$

**Solve for *n*. To check, substitute the answer and cross multiply.**

1. $\frac{9}{n} = \frac{6}{10}$

$$6 \times n = 9 \times 10$$
$$6n = 90$$
$$\frac{6n}{6} = \frac{90}{6}$$
$$n = 15$$
$$\text{Check: } \frac{9}{15} \times \frac{6}{10}$$
$$15 \times 6 = 9 \times 10$$
$$90 = 90$$

2. $\frac{12}{9} = \frac{8}{n}$

3. $\frac{2}{10} = \frac{n}{15}$

4. $\frac{n}{5} = \frac{12}{2}$

5. $\frac{90}{30} = \frac{60}{n}$

6. $\frac{2}{n} = \frac{3}{12}$

7. $\frac{n}{15} = \frac{8}{10}$

8. $\frac{15}{6} = \frac{n}{4}$

**Write a proportion and solve the word problems.**

9. A postal employee can sort 3 batches of mail in 10 minutes. How much time is needed to sort 15 batches?

10. A grocery store is selling 3 pounds of bananas for $.96. At this price, how much will 2 pounds of bananas cost?

**Check your answers on page 184.**

## Using Equivalent Fractions, Decimals, and Percents

You can find equivalent fractions, decimals, and percents.

| Fraction | Decimal | Percent |
|---|---|---|
| $\frac{1}{5}$ | 0.20 | 20% |
| $\frac{1}{2}$ | 0.50 | 50% |

**Strategy** Try one of these strategies with the following example.

| Fraction to Decimal: $\frac{1}{5}$ | | Decimal to Fraction: 0.34 | |
|---|---|---|---|
| Step 1 | Divide the numerator by the denominator. 1 ÷ 5 | Step 1 | Write the number without a decimal point as the numerator. 34 |
| Step 2 | Continue until the remainder is 0, or round to desired place. $5\overline{)1.00}$ = 0.20 | Step 2 | Use the place value of the last decimal digit as the denominator. $\frac{34}{100}$ |
| | | Step 3 | Reduce. $\frac{34}{100} = \frac{17}{50}$ |

| Decimal to Percent: 0.375 | | Percent to Decimal: 82% | |
|---|---|---|---|
| Step 1 | Move the decimal point two places to the right. 37.5 | Step 1 | Drop the %. Add a decimal point to the right of the ones digit. 82 |
| Step 2 | Add the percent sign. 37.5% | Step 2 | Move the decimal point two places to the left. 0.82 |

**Example**

Change $\frac{5}{8}$ to a decimal.

(1) 6.25
(2) .625
(3) 62.5
(4) .0625

You followed the steps for changing a fraction to a decimal. In Step 1 you changed $\frac{5}{8}$ to a decimal by dividing 5 by 8. In Step 2 you continued dividing to 0.625. The correct answer is choice (2).

## Practice

**Practice the strategy. Use the steps you learned. Solve the problem.**

**1.** What is the amount of ribbon written as a decimal?

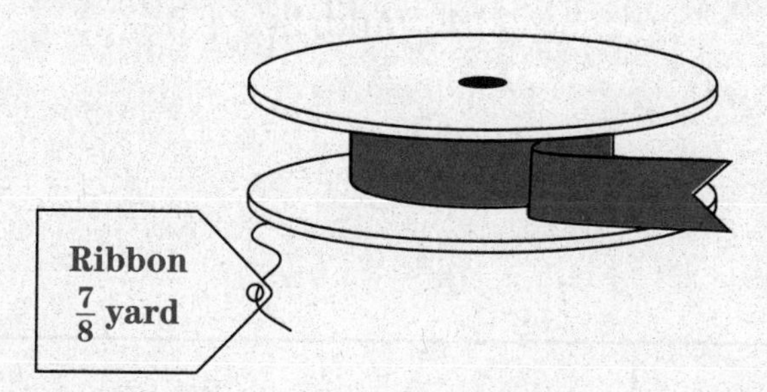


(1) 87.5
(2) 8.75
(3) 0.875
(4) 0.0875

**2.** Write the discount as a fraction.

(1) $\frac{1}{4}$
(2) $\frac{3}{8}$
(3) $\frac{1}{3}$
(4) $\frac{2}{3}$

**3.** Paul has $0.89 worth of change in his pocket. What fraction of a dollar is this?

(1) $\frac{8}{10}$
(2) $\frac{89}{100}$
(3) $\frac{100}{89}$
(4) $\frac{8}{9}$

**4.** Jade wrote 4.75 hours as the time needed to complete a job. How is this number written as a mixed number?

(1) $4\frac{1}{4}$
(2) $\frac{3}{4}$
(3) $4\frac{3}{4}$
(4) $3\frac{3}{4}$

**5.** The teacher reported that $\frac{3}{5}$ of the class handed in an extra-credit paper. What is this amount written as a percent?

(1) 6%
(2) 16%
(3) 50%
(4) 60%

**6.** A company pays 70% of the employees' health insurance premiums. What is this amount written as a decimal?

(1) 0.7
(2) 7.0
(3) 70.0
(4) .07

**Check your answers on page 185.**

# Unit 5 Wrap-up

Below are examples of the skills you learned in this unit. Read the examples and do the problems. Then check your answers.

**Examples.**

1. Add: $2\frac{1}{2} + 5\frac{3}{5}$

$$\begin{aligned} 2\tfrac{1}{2} &= 2\tfrac{5}{10} \\ +5\tfrac{3}{5} &= 5\tfrac{6}{10} \\ & \quad 7\tfrac{11}{10} = 8\tfrac{1}{10} \end{aligned}$$

2. Subtract: $\frac{4}{5} - \frac{2}{5}$

$$\frac{4}{5} - \frac{2}{5} = \frac{2}{5}$$

3. Multiply: $3\frac{1}{4} \times 1\frac{1}{2}$

$$3\frac{1}{4} \times 1\frac{1}{2} = \frac{13}{4} \times \frac{3}{2} = \frac{39}{8} = 4\frac{7}{8}$$

4. Divide: $5 \div 2\frac{1}{4}$

$$5 \div 2\frac{1}{4} = \frac{5}{1} \div \frac{9}{4} = \frac{5}{1} \times \frac{4}{9} = \frac{20}{9} = 2\frac{2}{9}$$

5. Change $\frac{4}{5}$ to a percent.

$$\frac{4}{5} = 4 \div 5 = 0.80 = 80\%$$

6. Solve for $n$. $\frac{24}{n} = \frac{36}{6}$

$$36 \times n = 24 \times 6$$

$$36n = 144$$

$$\frac{36n}{36} = \frac{144}{36}$$

$$n = 4$$

**Problems.**

1. Add: $2\frac{7}{10} + 5\frac{2}{3}$

2. Subtract: $\frac{7}{8} - \frac{3}{8}$

3. Multiply: $2\frac{2}{5} \times 3\frac{3}{4}$

4. Divide: $2\frac{3}{8} \div 2$

5. Change $\frac{1}{4}$ to a percent.

6. Solve for $n$. $\frac{15}{20} = \frac{n}{16}$

**Check your answers on page 185.**

# Unit 5 Practice

**Change each mixed number to an improper fraction.**

1. $2\frac{1}{6} =$
2. $3\frac{7}{8} =$
3. $5\frac{1}{3} =$
4. $1\frac{3}{5} =$

**Solve. Change improper fraction answers to mixed numbers.**
**Reduce answers if possible.**

5. $\frac{1}{5} + \frac{1}{5} =$
6. $\frac{3}{4} - \frac{1}{3} =$
7. $\frac{7}{8} \times \frac{3}{5} =$
8. $\frac{7}{12} \div \frac{1}{3} =$

**Solve. Reduce answers if possible.**

9. $4\frac{1}{6} + 2\frac{1}{3} =$
10. $12\frac{4}{5} - 1\frac{2}{3} =$
11. $2\frac{1}{4} \times 4 =$
12. $3\frac{1}{4} \div 2 =$

**Reduce each ratio to lowest terms.**

13. $\frac{9}{6}$
14. $\frac{10}{15}$
15. $\frac{40}{20}$
16. $\frac{3}{12}$
17. $\frac{4}{10}$
18. $\frac{75}{25}$

**Solve for *n*.**

19. $\frac{4}{n} = \frac{20}{55}$
20. $\frac{n}{20} = \frac{5}{25}$
21. $\frac{9}{6} = \frac{n}{8}$
22. $\frac{32}{12} = \frac{24}{n}$

**Solve each problem. Round percents to the nearest tenth.**

23. Change 20% to a fraction.
24. Change 105% to a mixed number.
25. Change $\frac{11}{20}$ to a percent.

**Check your answers on page 185.**

# Unit 5 GED Test Practice

**Read each problem carefully. Circle the number of the correct answer.**

1. Marina stated that $\frac{4}{16}$ of the people on her team wear glasses. What is this fraction in lowest terms?

   (1) $\frac{1}{6}$
   (2) $\frac{1}{4}$
   (3) $\frac{1}{3}$
   (4) $\frac{1}{2}$

2. The table shows the distances four workers live from their office. What is the order of these fractions from least to greatest?

   $\frac{3}{10}$ mile $\quad \frac{2}{5}$ mile $\quad \frac{1}{3}$ mile $\quad \frac{5}{6}$ mile

   (1) $\frac{5}{6}, \frac{2}{5}, \frac{1}{3}, \frac{3}{10}$
   (2) $\frac{3}{10}, \frac{2}{5}, \frac{1}{3}, \frac{5}{6}$
   (3) $\frac{1}{3}, \frac{3}{10}, \frac{2}{5}, \frac{5}{6}$
   (4) $\frac{3}{10}, \frac{1}{3}, \frac{2}{5}, \frac{5}{6}$

3. Leah worked for $3\frac{1}{2}$ hours on Saturday and $4\frac{1}{3}$ hours on Sunday. How many hours did she work in all?

   (1) $8\frac{5}{6}$
   (2) $7\frac{5}{6}$
   (3) 7
   (4) $1\frac{1}{6}$

4. A container has $\frac{3}{4}$ cup of oil. Leo uses $\frac{1}{4}$ cup of the oil to make a cake. How much oil is left in the container?

   (1) $\frac{3}{16}$ cup
   (2) $\frac{3}{8}$ cup
   (3) $\frac{1}{2}$ cup
   (4) 1 cup

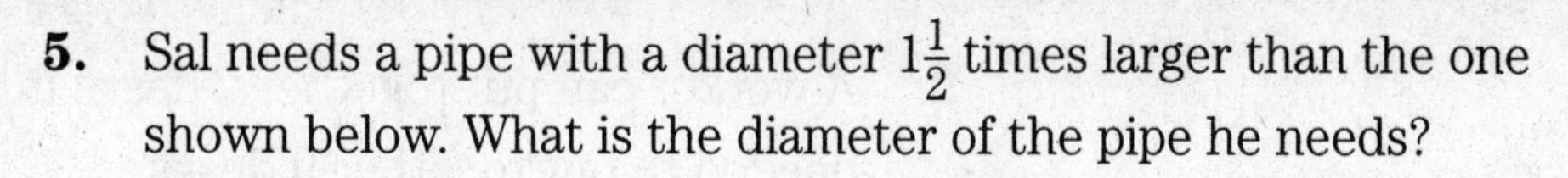

**5.** Sal needs a pipe with a diameter $1\frac{1}{2}$ times larger than the one shown below. What is the diameter of the pipe he needs?

(1) $2\frac{1}{6}$ inches
(2) $3\frac{1}{2}$ inches
(3) $4\frac{3}{4}$ inches
(4) $4\frac{7}{8}$ inches

**6.** Carole has a $\frac{3}{4}$-pound block of chocolate. She cuts chunks that are $\frac{1}{16}$-pound each. How many chunks of chocolate does she cut?

(1) $\frac{3}{64}$
(2) $\frac{3}{16}$
(3) 6
(4) 12

**7.** This year, wireless telephone sales increased by 220%. How is this percent written as a mixed number?

(1) $2\frac{1}{20}$
(2) $2\frac{1}{5}$
(3) $2\frac{2}{5}$
(4) $22\frac{1}{10}$

**8.** There were 80 people registered for a seminar. Of the 80 people registered, only 56 people showed up on the day of the seminar. What is the ratio of people who attended to people who registered?

(1) $\frac{10}{7}$
(2) $\frac{8}{7}$
(3) $\frac{7}{10}$
(4) $\frac{8}{10}$

9. A worker can put together 2 desks in 5 hours. How much time is needed to put together 9 desks?

(1) 90 hours
(2) 45 hours
(3) 22.5 hours
(4) 3.6 hours

10. Of the people surveyed, 72% said they would vote for Ellis as mayor. How is this percent written as a decimal?

(1) 72.0
(2) 7.2
(3) 0.72
(4) 0.072

**Check your answers on page 186.**

## Unit 5 Skill Check-Up Chart

Check your answers. In the first column, circle the numbers of questions that you missed. Then look across the rows to see the skills you need to review and the pages where you can find each skill.

| Question | Skill | Page |
|---|---|---|
| 1 | Reducing Fractions | 114 |
| 2 | Common Denominators | 116 |
| 3 | Adding Mixed Numbers with Unlike Denominators | 123 |
| 4 | Subtracting Fractions with Like Denominators | 124 |
| 5 | Multiplying Mixed Numbers and Whole Numbers | 131 |
| 6 | Dividing Fractions by Fractions | 132 |
| 7 | Percents as Fractions | 134 |
| 8 | Ratios | 135 |
| 9 | Solving Proportions | 137 |
| 10 | Using Equivalent Fractions, Decimals, and Percents | 138–139 |

# Mathematics Posttest

This *Posttest* will give you an idea of how well you have learned the skills that are in this book. You will solve each problem and choose the correct answer. There is no time limit.

**Read each problem carefully. Circle the number of the correct answer.**

**1.** There were 10,641 fans at a basketball game. Rounded to the lead digit, how many fans is this?

(1) 10,000
(2) 10,600
(3) 10,640
(4) 11,000

**2.** Tara has 12 boxes of beads like those shown below. How many beads does Tara have in all?

(1) 40
(2) 468
(3) 492
(4) 5,760

**3.** At the beginning of the year, a volunteer group had 19 members. During the year 13 more members joined. How many members are now in the group?

(1) 33
(2) 32
(3) 22
(4) 6

**4.** Rachel spent $175 to rent a car for 5 days. The cost each day was the same. How much was the daily cost?

(1) $875
(2) $180
(3) $35
(4) $25

**5.** Ellen earned $1,920 last month. Her monthly bills totaled $1,375. How much does Ellen have left after paying her bills for the month?

(1) $3,295
(2) $655
(3) $555
(4) $545

**6.** There are 6,947 residents in the village of Alton. What is the value of the digit 6 in this number?

(1) 6 ones
(2) 6 tens
(3) 6 hundreds
(4) 6 thousands

7. Rounded to the nearest ten, what is the number of graduates?

(1) 1,300
(2) 1,290
(3) 1,280
(4) 1,000

8. Last year an accounting firm prepared 1,319 business tax returns and 1,561 personal tax returns. Which item correctly compares the two numbers?

(1) $1,319 < 1,561$
(2) $1,319 > 1,561$
(3) $1,561 < 1,319$
(4) $1,319 = 1,561$

9. A shipment contains 5 packages, each weighing 14 pounds. What is the total weight of all packages in the shipment?

(1) 9 pounds
(2) 19 pounds
(3) 50 pounds
(4) 70 pounds

10. If Andy paid $60 for four sessions of golf, how much did each session of golf cost?

(1) $60
(2) $56
(3) $15
(4) $12

11. Kate changed her wireless phone plan and got a new phone. The price of the phone was $99, and she got a $40 rebate. How much did Kate pay for the phone after the rebate?

(1) $139
(2) $99
(3) $59
(4) $40

12. A company with 556 workers just hired another 192 workers. How many people now work for the company?

(1) 748
(2) 556
(3) 364
(4) 192

**13.** Tim rode his bicycle for 13.8 miles on Saturday. On Sunday, he biked 7.6 miles. How many miles did Tim ride in all that weekend?

(1) 6.2
(2) 20.0
(3) 20.4
(4) 21.4

**14.** Below is Rhonda's pay stub. The deductions are taken from the gross pay to find the net pay. What will Rhonda's net pay be?

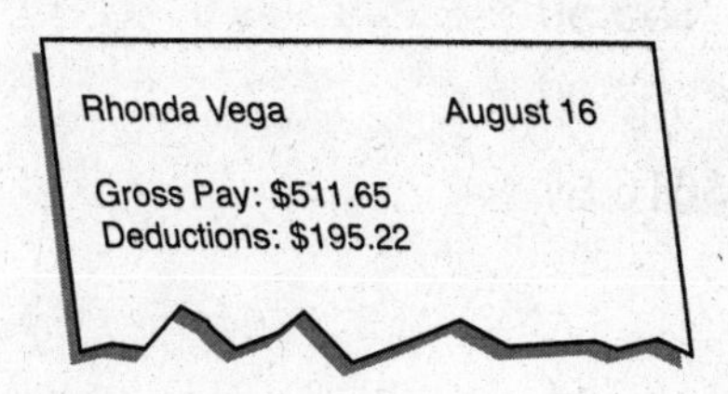


(1) $316.43
(2) $416.43
(3) $426.43
(4) $706.87

**15.** Jules and his brother shared a pizza at a restaurant. The pizza and drinks cost a total of $14.90. If the brothers share the bill equally, how much will each brother pay?

(1) $29.80
(2) $12.90
(3) $7.45
(4) $7.00

**16.** Each week $26.50 out of Rhonda's pay goes into a retirement account. Rhonda worked 48 weeks last year. How much of Rhonda's pay went into her retirement account last year?

(1) $74.50
(2) $127.20
(3) $1,008.00
(4) $1,272.00

**17.** Tran bought some meat and cheese to make a deli platter. His receipt is shown below. Round to the lead digit to estimate the total weight of the meat and cheese.

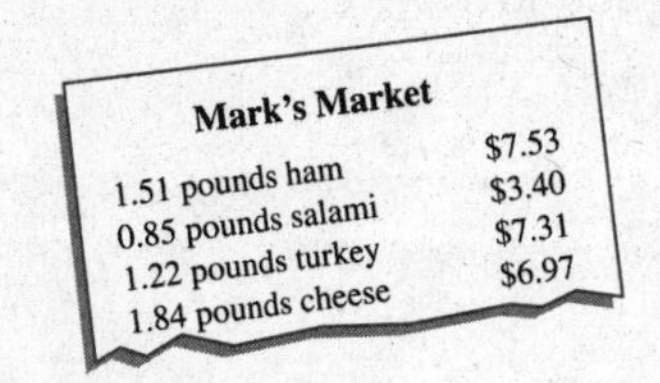


(1) 3 pounds
(2) 4 pounds
(3) 6 pounds
(4) 7 pounds

**18.** How many bags measuring $\frac{1}{12}$ cubic yard can be made from $2\frac{2}{3}$ cubic yards of mulch?

(1) $\frac{2}{9}$
(2) $2\frac{7}{12}$
(3) $2\frac{3}{4}$
(4) 32

**19.** Leo's salary is $1,850 per month. He is getting a 3% raise. What is the amount of the raise per month?

(1) $5.55
(2) $55.50
(3) $555.00
(4) $616.67

**20.** Jamal grew $1\frac{1}{4}$ inches in one year and $2\frac{1}{2}$ inches in the next year. How much did Jamal grow in these 2 years?

(1) $1\frac{1}{4}$ inches
(2) $3\frac{1}{8}$ inches
(3) $3\frac{1}{3}$ inches
(4) $3\frac{3}{4}$ inches

**21.** During a storm, $3\frac{7}{10}$ inches of rain fell on Monday and $1\frac{3}{10}$ inches fell on Tuesday. How much more rain fell on Monday?

(1) $2\frac{2}{5}$ inches
(2) $2\frac{7}{10}$ inches
(3) $4\frac{1}{2}$ inches
(4) 5 inches

**22.** At work it takes James 1 minute to pack 12 bottles of shampoo into boxes. At this rate, how many bottles can he pack in 1 hour (60 minutes)?

(1) 720
(2) 360
(3) 72
(4) 5

**This circle graph shows the results of a survey. Use the graph for problems 23 and 24.**

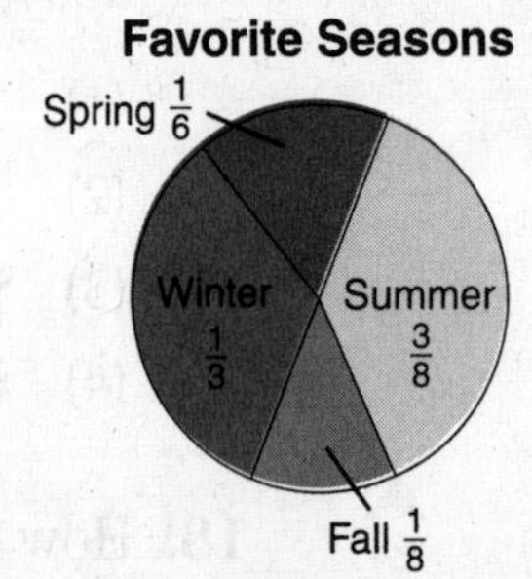


**23.** Men made up $\frac{1}{3}$ of the people who chose spring as their favorite season. What fraction of the entire group are men who chose spring as their favorite season?

(1) 2
(2) $\frac{1}{2}$
(3) $\frac{1}{9}$
(4) $\frac{1}{18}$

**24.** Which fraction is equivalent to the fraction of the people who chose summer as their favorite season?

(1) $\frac{8}{24}$
(2) $\frac{9}{24}$
(3) $\frac{8}{16}$
(4) $\frac{12}{24}$

**When you finish the *Mathematics Posttest*, check your answers on page 186. Then look at the chart on page 149.**

# Skills Review Chart

This chart shows you which skills you should review. Check your answers. In the first column, circle the number of any questions you missed. Then look across the row to find out which skills you should review as well as the page numbers on which you find instruction on those skills. Compare the items you circled in the *Skills Review Chart* to those you circled in the *Skills Preview Chart* to see the progress you've made.

| Questions | Skill | Pages |
|---|---|---|
| 6, 8 | Whole Number Place Value | 10–17 |
| 3, 12 | Whole Number Addition | 28–31 |
| 5, 11 | Whole Number Subtraction | 34–37 |
| 2, 9 | Whole Number Multiplication | 54–59 |
| 4, 10 | Whole Number Division | 60–64 |
| 1, 7 | Whole Number Rounding and Estimation | 44–45 |
| 13 | Decimal Addition | 84 |
| 14 | Decimal Subtraction | 85 |
| 16 | Decimal Multiplication | 86–87 |
| 15 | Decimal Division | 88–90 |
| 17 | Decimal Rounding and Estimation | 82–83 |
| 19 | Percents | 96–97 |
| 24 | Equivalent Fractions | 113–115 |
| 20 | Fraction Addition | 120–123 |
| 21 | Fraction Subtraction | 124–127 |
| 23 | Fraction Multiplication | 130–131 |
| 18 | Fraction Division | 132–133 |
| 22 | Ratio and Proportion | 135–137 |

# Glossary

**common denominator** A number that two or more denominators divide into evenly. *page 116*

**decimal** A number that shows an amount less than 1. *page 78*

**denominator** The bottom number in a fraction; tells how many equal parts are in the whole. *page 112*

**digit** One of ten symbols—0, 1, 2, 3, 4, 5, 6, 7, 8, 9—used to write numbers. *page 10*

**dividend** The number that is being divided; in a fraction, the dividend is the numerator. *page 60*

**divisor** The number that is used to divide another number; in a fraction, the divisor is the denominator. *page 60*

**equal ratios** Equal fractions. *page 136*

**equivalent fractions** Different fractions such as $\frac{1}{2}$ and $\frac{2}{4}$ that represent the same amount. *page 113*

**estimate** To find an answer by rounding the numbers in a problem. *page 44*

**fraction bar** The line that separates the numerator and denominator. *page 112*

**improper fraction** A fraction with a numerator that is greater than or equal to the denominator. *page 118*

**invert** To turn a fraction upside down. *page 132*

**lead digit** The first digit on the left in a number with two or more digits. *page 44*

**lowest common denominator (LCD)** The smallest number that two or more denominators can divide into evenly. *page 117*

**mixed number** A number made up of a whole number and a fraction part. *page 118*

**numerator** The top number in a fraction; tells how many parts of the whole or group are being considered. *page 112*

**operation** The process you use to solve a math problem. Addition, subtraction, multiplication, and division are the basic mathematical operations. *page 70*

**partial product** The total you get when multiplying a number by one digit of another number. *page 56*

**percent** Hundredths; when using percents, the whole is divided into 100 equal parts. *page 94*

**place name** The name of the place—ones, tens, hundreds, etc.—held by the digits in a number. *page 10*

**place value** The value of a digit based on its position in a number. *page 10*

**proper fraction** A fraction with a numerator that is less than the denominator. *page 118*

**proportion** Two equal ratios. *page 137*

**quotient** The answer to a division problem. *page 60*

**ratio** A fraction that shows a relationship between two numbers. *page 135*

**reducing** Reducing a fraction to lowest terms means dividing both the numerator and denominator by the same number. *page 114*

**remainder** The amount left over in a division problem. *page 62*

**rename** Regrouping numbers using different place values. *page 14*

**round** An approximation to the nearest number when an exact number is not necessary. Rounding is expressing a number to the nearest ten, hundred, or thousand. *page 17*

# Answers and Explanations

## Mathematics Pretest, *pages 3–7*

### Page 3

**1.** 4 thousands

**2.** 6 hundreds

**3.** 2 hundred thousands

**4.** 54

**5.** 318

**6.** 2,709

**7.** >

**8.** <

**9.** >

**10.** 660

**11.** 340

**12.** 110

**13.** Elizabeth

**14.** 590 people

### Page 4

**15.** 13

**16.** 9

**17.** 13

**18.** $8 + 7 = 15$

**19.** $16 - 9 = 7$

**20.** $69

**21.** 30

**22.** $227

**23.** 508

**24.** $698

**25.** 3,143

**26.** 110

**27.** 100

**28.** 1,400

**29.** 71 people

**30.** 238 people

### Page 5

**31.** $369

**32.** 6,500

**33.** 64,538

**34.** 39

**35.** 41

**36.** 51

**37.** 2,400

**38.** 400

**39.** 20,000

**40.** $4

**41.** 168

### Page 6

**42.** 5.9

**43.** 123.5

**44.** 1.5

**45.** 20.3

**46.** $26.43

**47.** 6.01

**48.** 13.984

**49.** $3.66

**50.** 5.4

**51.** 22.01

**52.** 9.9

**53.** 12.9

**54.** 12

**55.** 175%

**56.** 40%

**57.** $0.12

**58.** $58.50 per week

### Page 7

**59.** $\frac{1}{4}$

**60.** $\frac{20}{25}$

**61.** $\frac{21}{24}$

**62.** $\frac{2}{15}$

**63.** $\frac{1}{2}$

**64.** $3\frac{13}{28}$

**65.** $\frac{13}{30}$

**66.** $5\frac{19}{24}$

**67.** $\frac{1}{12}$

**68.** $\frac{22}{27}$

**69.** 11

**70.** 3

**71.** $n = 7$

**72.** $n = 9$

**73.** $n = 5$

**74.** $n = 3$

**75.** $17\frac{1}{2}$ miles

**76.** 6 servings

## Unit 1

### Page 10

| | | | |
|---|---|---|---|
| **1.** | | 6 | 4 |
| **2.** | 7 | 3 | 9 |
| **3.** | | 8 | 7 |
| **4.** | 4 | 0 | 6 |

**5.** tens

**6.** hundreds

**7.** tens

**8.** hundreds

**9.** ones

**10.** tens

**11.** tens

**12.** ones

**13.** ones

**14.** hundreds

**15.** ones

**16.** ones

## Page 11

**17.** 2 hundreds

**18.** 5 ones; the model shows 5 groups of ones, or 5 ones.

**19.** 4 tens **20.** 2 tens

**21.** 7 ones; the 7 is in the ones place and has a value of 7.

**22.** 2 hundreds **23.** 9 ones

**24.** 9 hundreds **25.** 4 tens

**26.** 0 tens **27.** 1 hundred

**28.** 7 ones **29.** 6 tens

**30.** 0 ones **31.** 5 hundreds

## Page 12

**1.** 31

**2.** 204; 2 hundreds 4 ones = 200 + 4

**3.** 9 **4.** 48

**5.** 773 **6.** 50

**7.** 94 **8.** 686

**9.** 70

**10.** 8 tens
5 ones

**11.** 3 hundreds
2 tens
1 one

**12.** 8 hundreds
0 tens
3 ones

**13.** 2 tens
0 ones

**14.** 1 hundred
1 ten
6 ones

**15.** 2 hundreds
0 tens
0 ones

## Page 13

**1.** 121

**2.** 4 tens + 8 ones = 48

**3.** 105 **4.** 20

**5.** yes
3 hundreds 4 tens 6 ones = 346

**6.** 735 stamps
7 hundreds 3 tens 5 ones = 735

## Page 14

**1.** 30

**2.** 4; You can rename 40 ones as 4 tens.

**3.** 100 **4.** 1 **5.** 1

**6.** 10 **7.** 24 **8.** 3; 9

**9.** 146 **10.** 21

**11.** 340 sheets
3 hundreds 4 tens = 340

**12.** 240 paper clips
2 hundreds 4 tens = 240

## Page 15

**1.** 4 400,000
5 50,000
9 9,000
3 300
2 20
8 8

**2.** 4 40,000
7 7,000
0 0
0 0
6 6

| | | | | | | |
|---|---|---|---|---|---|---|
| **1.** | 4 | 5 | 9, | 3 | 2 | 8 |
| **2.** | | 4 | 7, | 0 | 0 | 6 |

**3.** 400 **4.** 60,000

**5.** 7,000 **6.** 0

**7.** 0 **8.** 400,000

## Page 16

**1.** >

**2.** <

47 is less than 463 because 47 has 0 hundreds.

3. <  4. >  5. <
6. >  7. >  8. <
9. <  10. <  11. >
12. <  13. <  14. <
15. >  16. >  17. >
18. >  19. >  20. >
21. <  22. >  23. <
24. <

**Page 17**

1. 30
2. 50
   45 is exactly halfway between 40 and 50; round up.
3. 60
4. 40
5. 640
6. 670
7. 650
8. 690
9. exact answer
10. rounded answer

## GED Skill Strategy

**Page 19**

**Exercise 1**

1. 29.
2. 307.
3. 6280.
4. 60280.
5. 889002.
6. 223468.

**Exercise 2**

7. c  8. e
9. d  10. b
11. a

**Exercise 3**

12. 98,077  13. 76,222
14. 4,600  15. 23,756
16. 39,641  17. 1,842
18. 594,437  19. 9,046

## GED Test-Taking Strategy

**Page 21**

1. (2) tens
2. (1) 450
3. (3) 36
4. (2) $30
5. (3) hundreds

**Page 22**

1. 9 hundreds = 900
   0 tens = 0
   2 ones = 2
2. 4 ones
3. 6 hundreds
   5 tens
   8 ones
4. 572
5. 7,000
6. 37 < 73
7. 130

**Page 23**

**1.** 8 tens
**2.** 4 thousands
**3.** 41
**4.** 920
**5.** 765
**6.** 5
**7.** 3 thousands
**8.** 100 ones
**9.** 2 hundreds
**10.** 5 tens
**11.** $<$
**12.** $>$
**13.** $>$
**14.** $>$
**15.** $<$
**16.** $>$
**17.** $>$
**18.** $<$
**19.** 80
**20.** 70
**21.** 80
**22.** 90
**23.** 320
**24.** 360
**25.** 310
**26.** 330
**27.** Sharise
**28.** 30 hours

## GED Test Practice, *pages 24–26*

**Page 24**

1. **(30) 30,000** Choices (1), (2), and (4) are incorrect because the digit 3 is located in the ten thousands place.
2. **(3) 520** Choices (1), (2), and (4) are incorrect since the ones digit of 5 rounds up to the next ten (520).
3. **(4) 246** Choices (1), (2), and (3) are incorrect because the model shows 2 hundreds, 4 tens, and 6 ones.
4. **(2) tens** Choices (1), (3), and (4) are incorrect because the digit 4 in 142 is in the tens place.

**Page 25**

5. **(1) 250** Choices (2), (3), and (4) are incorrect because there are 2 hundreds and 5 tens, which is 250.
6. **(3) 22 tens** Choices (1), (2), and (4) are incorrect because 220 is equal to 22 tens.
7. **(3) 700** Choices (1), (2), and (4) are incorrect because the 7 is in the hundreds place, so the value is 700.
8. **(2) 70** Choices (1), (3), and (4) are incorrect because the ten closest to 67 is 70.

**Page 26**

9. **(1) $250** Choice (3) is incorrect because $247 should be rounded up to $250. Choices (2) and (4) are incorrect because they are not rounded to the nearest ten.
10. **(2) ten thousands** Choices (1), (3), and (4) are incorrect because the 3 in 38,942 is in the ten thousands place.

## Unit 2

**Page 28**

| + | 0 | 1 | 2 | 3 | 4 | 5 | 6 | 7 | 8 | 9 |
|---|---|---|---|---|---|---|---|---|---|---|
| 0 | 0 | 1 | 2 | 3 | 4 | 5 | 6 | 7 | 8 | 9 |
| 1 | 1 | 2 | 3 | 4 | 5 | 6 | 7 | 8 | 9 | 10 |
| 2 | 2 | 3 | 4 | 5 | 6 | 7 | 8 | 9 | 10 | 11 |
| 3 | 3 | 4 | 5 | 6 | 7 | 8 | 9 | 10 | 11 | 12 |
| 4 | 4 | 5 | 6 | 7 | 8 | 9 | 10 | 11 | 12 | 13 |
| 5 | 5 | 6 | 7 | 8 | 9 | 10 | 11 | 12 | 13 | 14 |
| 6 | 6 | 7 | 8 | 9 | 10 | 11 | 12 | 13 | 14 | 15 |
| 7 | 7 | 8 | 9 | 10 | 11 | 12 | 13 | 14 | 15 | 16 |
| 8 | 8 | 9 | 10 | 11 | 12 | 13 | 14 | 15 | 16 | 17 |
| 9 | 9 | 10 | 11 | 12 | 13 | 14 | 15 | 16 | 17 | 18 |

1. 10
2. 9

$$\begin{array}{r} 7 \\ +\,2 \\ \hline 9 \end{array}$$

3. 10
4. 5
5. 11
6. 9
7. 16
8. 4
9. 14
10. 18
11. 17
12. 13
13. 12
14. 14
15. 2
16. 12
17. 12
18. 10
19. 15
20. 11

$$\begin{array}{r} 7 \\ +\,4 \\ \hline 11 \end{array}$$

21. 10
22. 7

**Page 29**

1. 98
2. 99
3. 66
4. 55
5. 57
6. 369
7. 599
8. 978
9. 367
10. 866
11. 535
12. 756
13. 488
14. 843
15. 685
16. 1,876
17. 4,878
18. 7,766
19. 8,757
20. 4,053
21. 5,550
22. 5,088
23. 8,189

**Page 30**

1. 151
2. 74

Add the ones. 8 + 6 = 14. Carry 1 ten.

Add the tens. 1 + 6 = 7.

$$\begin{array}{r} {}^{1} \\ 68 \\ +\,6 \\ \hline 74 \end{array}$$

3. 61
4. 80
5. 127
6. 100
7. 23
8. 80
9. 110
10. 51
11. 614
12. 724
13. 1,031
14. 1,163
15. 717
16. 2,098
17. 8,105
18. 4,700
19. 11,520

**Page 31**

1. 50
2. 13

Add the first two digits. 5 + 2 = 7.

Then add the last digit to the 7.

7 + 6 = 13

$$\begin{array}{r} 5 \\ 2 \\ +\,6 \\ \hline 13 \end{array}$$

3. 111
4. 634
5. 353
6. 895
7. 788
8. 587
9. 832
10. 1,120
11. 180 gallons

$$\begin{array}{r} {}^{12} \\ 38 \\ 75 \\ +\,67 \\ \hline 180 \end{array}$$

12. 318 tickets

$$\begin{array}{r} {}^{21} \\ 192 \\ 67 \\ +\;\;59 \\ \hline 318 \end{array}$$

## GED Skill Strategy

**Page 32**

1. 56.
2. 56.
3. 48.
4. 104.
5. 189.
6. 293.

7. 55.

8. 348.

9. 237.

10. 237.

11. 663.

12. 900.

13. 1145.

14. 2045.

15. 590.

16. 2635.

**Page 33**

**17.** 79

**18.** 47

**19.** 127

**20.** 129

**21.** 684

**22.** 1,035

**23.** 1,681

**24.** 1,028

**25.** 157

**26.** 930

**27.** 4,265

**28.** 4,101

**29.** $214

**30.** 544 miles

**31.** 461 days

**32.** 135 items

**Page 34**

**1.** 6

**2.** 4

**3.** 3

**4.** 8

**5.** 7

**6.** 5

**7.** 7

**8.** 9

**9.** 0

**10.** 2

**11.** 9

**12.** 9

**13.** 6

**14.** 7

**15.** 4

**16.** 3

**17.** 4

**18.** 7

**Page 35**

**1.** 24

**2.** 37

Subtract the digits in the ones column. $7 - 0 = 7$. Then subtract the digits in the tens column. $6 - 3 = 3$. Add to check.

$$\begin{array}{r} 67 \\ -\,30 \\ \hline 37 \end{array} \qquad \begin{array}{r} 37 \\ +\,30 \\ \hline 67 \end{array}$$

**3.** 31

**4.** 84

**5.** 21

**6.** 61

**7.** 12

**8.** 11

**9.** 612

**10.** 120

**11.** 312

**12.** 291

**13.** 3,147

**14.** 1,123

**15.** 3,262

**16.** 4,423

**Page 36**

**1.** 18

**2.** 19

Since you can't subtract 7 from 6, borrow 1 ten. Rename the borrowed ten as 10 ones and add it to the 6 ones. $16 - 7 = 9$ ones. Then subtract the tens. $3 - 2 = 1$ ten. Add to check.

$$\begin{array}{r} {\scriptstyle 3\,16} \\ \not{4}\not{6} \\ -\,27 \\ \hline 19 \end{array} \qquad \begin{array}{r} {\scriptstyle 1} \\ 19 \\ +\,27 \\ \hline 46 \end{array}$$

**3.** 67

**4.** 28

**5.** 5

**6.** 25

**7.** 6

**8.** 86

**9.** 385

**10.** 107

**11.** 791

**12.** 206

**13.** 1,182

**14.** 1,557

**15.** 4,398

**16.** 5,341

## Page 37

**1.** 135

**2.** 664

Borrow 1 hundred across the zero from the hundreds column. Rename 1 hundred as 10 tens. Borrow 1 ten and rename as 10 ones. You now have 6 hundreds, 9 tens, and 13 ones. Subtract.

13 – 9 = 4 ones

9 – 3 = 6 tens

6 – 0 = 6 hundreds

Add to check

$$\begin{array}{r} {\scriptstyle 6\,9\,13} \\ \not{7}\not{0}\not{3} \\ -\ 39 \\ \hline 664 \end{array} \qquad \begin{array}{r} {\scriptstyle 1} \\ 664 \\ +\ 39 \\ \hline 703 \end{array}$$

**3.** 503

**4.** 748

**5.** 367

**6.** 325

**7.** 582

**8.** 79

**9.** 42

**10.** 376

**11.** 98

**12.** 819

**13.** 3,237

**14.** 2,929

**15.** 645

**16.** 6,544

## Page 38

**1.** 84.

**2.** 84.

**3.** 39.

**4.** 45.

**5.** 19.

**6.** 26.

**7.** 1076.

**8.** 1076.

**9.** 539.

**10.** 537.

**11.** 250.

**12.** 287.

## Page 39

**13.** 34

**14.** 17

**15.** 26

**16.** 12

**17.** $49

**18.** 27

**19.** 37

**20.** 329

**21.** 219

**22.** 141

**23.** $39

**24.** 581

**25.** $513

**26.** 651

**27.** $65

**28.** $17

**29.** 48 days

**30.** 54 stamps

**Page 40**

**1.** 372

**2.** 125

First put the digits in columns. Add the ones. 7 + 8 = 15. Carry 1 ten. Add the tens. 1 + 3 + 8 = 12.

$$\begin{array}{r} {}^{1}\phantom{0} \\ 37 \\ +\ 88 \\ \hline 125 \end{array}$$

**3.** 81

**4.** 54

**5.** 1,480

**6.** 4,060

**7.** 4,072

**8.** 14,018

**9.** 309

**10.** 90

**11.** 463

**12.** 4,104

**13.** $153

$$\begin{array}{r} {}^{1}\phantom{00} \\ \$125 \\ +\ \ 28 \\ \hline \$153 \end{array}$$

**14.** $203

$$\begin{array}{r} {}^{1\,1}\phantom{0} \\ \$135 \\ +\ \ 68 \\ \hline \$203 \end{array}$$

**Page 41**

**1.** 42

**2.** 33

Line up the digits. Subtract the digits in the ones column. 6 – 3 = 3. Then subtract the digits in the tens column. 5 – 2 = 3. Add to check.

$$\begin{array}{r} 56 \\ -\ 23 \\ \hline 33 \end{array} \qquad \begin{array}{r} 33 \\ +\ 23 \\ \hline 56 \end{array}$$

**3.** 51

**4.** $52

**5.** 345

**6.** $643

**7.** 565

**8.** 568

**9.** 8,212

**10.** $4,120

**11.** $386

**12.** 2,378

**13.** $345

$$\begin{array}{r} \$865 \\ -\ 520 \\ \hline \$345 \end{array}$$

**14.** 313 miles

$$\begin{array}{r} 638 \\ -\ 325 \\ \hline 313 \end{array}$$

**Page 42**

**1.** 33

**2.** 80

Subtract the first two numbers. 76 – 19 = 57. Then add the answer to the third number. 57 + 23 = 80.

$$\begin{array}{r} {}^{6\,16} \\ 7\not{6} \\ -\,19 \\ \hline 57 \end{array} \qquad \begin{array}{r} {}^{1}\phantom{0} \\ 57 \\ +\,23 \\ \hline 80 \end{array}$$

**3.** 27

**4.** 44

**5.** 249

**6.** $425

**7.** $518

**8.** $60

**9.** 81

**10.** 38

**11.** 325

**12.** $873

**13.** 982

**14.** $3

$13 + $4 = $17

$20 – $17 = $3

**15.** 1,348 empty seats

1,287 + 365 = 1,652

3,000 – 1,652 = 1,348

### Page 43

**1.** 197

**2.** 169

Subtract the first two numbers.

256 − 73 = 183. Subtract the third number from the answer. 183 − 14 = 169.

$$\begin{array}{r} \scriptstyle 1\,15 \\ 2\not{5}6 \\ -\ 73 \\ \hline 183 \end{array} \qquad \begin{array}{r} \scriptstyle 7\,13 \\ 1\not{8}\not{3} \\ -\ 14 \\ \hline 169 \end{array}$$

**3.** 110 **4.** 110

**5.** 705 **6.** 757

**7.** 167 **8.** 413

**9.** 656 **10.** 250

**11.** $9

$30 − $4 = $26

$26 − $17 = $9

**12.** $19

$20 − $7 = $13

$13 + $5 + $1 = $19

## GED Skill Strategy

### Page 44

**1.** 400 **2.** 90

**3.** 2,000 **4.** 800

**5.** 30 **6.** 700

**7.** 4,000 **8.** 600

### Page 45

**9.** 30

**10.** 120

**11.** 120

**12.** 220

**13.** $470

**14.** 870

**15.** 510

**16.** 400

**17.** 800

**18.** 1,100

**19.** $7,200

**20.** 2,000

**21.** 5,500

**22.** 7,000

**23.** $1,000

**24.** $800

**25.** 6,000 miles

**26.** 60 cars

## GED Test-Taking Strategy

### Page 47

**1.** **(2) $305** Choice (2) is correct because $165 plus $140 is $305.

**2.** **(4) 116** Choice (4) is correct because 128 ounces minus 12 ounces is 116 ounces.

**3.** **(3) 219** Choice (3) is correct because 48 plus 67 plus 104 is 219 customers.

**4.** **(1) $123** Choice (1) is correct because $542 minus $419 is $123.

**5.** **(3) 154** Choice (3) is correct because 432 phones minus 278 phones is 154 phones.

**6.** **(4) 187** Choice (4) is correct because 78 miles plus 109 miles is 187 miles.

### Page 48

**1.** 6,035 **2.** 3,921

**3.** 604 **4.** 4,125

**5.** 443 **6.** 3,770

## Page 49

1. 145
2. 82
3. 895
4. 4,346
5. 4,636
6. 77
7. 171
8. 857
9. 2,702
10. 52
11. 46
12. 847
13. 454
14. 3,753
15. 53
16. 54
17. 253
18. 7,343
19. 279
20. 463
21. 6,861
22. 3,868
23. 6,425
24. 378 reams of white paper

$$\begin{array}{r} 456 \\ -\ 78 \\ \hline 378 \end{array}$$

25. $448

$$\begin{array}{r} \$160 \\ +\ 288 \\ \hline \$448 \end{array}$$

## GED Test Practice, *pages 50–52*

### Page 50

1. **(1) 138 miles** Choice (1) is correct because 6,000 miles – 5,862 miles = 138 miles.
2. **(2) $259** Choice (2) is correct because $297 – $38 = $259.
3. **(1) 265 people** Choice (1) is correct because 136 workers + 129 workers = 265.
4. **(2) 43 + 13** Choice (2) is correct because operations in parentheses come first.

### Page 51

5. **(3) 340 boxes** Choice (3) is correct because 120 boxes + 180 boxes + 40 boxes = 340 boxes.
6. **(4) 827 square feet** Choice (4) is correct because 1,200 square feet – 325 square feet – 48 square feet = 827 square feet.
7. **(1) $378** Choice (1) is correct because $420 – $42 = $378.
8. **(3) 70 square yards** Choice (3) is correct because 50 + 20 = 70.

### Page 52

9. **(3) 335 tons** Choice (3) is correct because 578 tons – 243 tons = 335 tons.
10. **(4) 76 necklaces** Choice (4) is correct because 83 necklaces – 15 necklaces + 8 necklaces = 76 necklaces.

## Unit 3

### Page 54

| × | 0 | 1 | 2 | 3 | 4 | 5 | 6 | 7 | 8 | 9 |
|---|---|---|---|---|---|---|---|---|---|---|
| 0 | 0 | 0 | 0 | 0 | 0 | 0 | 0 | 0 | 0 | 0 |
| 1 | 0 | 1 | 2 | 3 | 4 | 5 | 6 | 7 | 8 | 9 |
| 2 | 0 | 2 | 4 | 6 | 8 | 10 | 12 | 14 | 16 | 18 |
| 3 | 0 | 3 | 6 | 9 | 12 | 15 | 18 | 21 | 24 | 27 |
| 4 | 0 | 4 | 8 | 12 | 16 | 20 | 24 | 28 | 32 | 36 |
| 5 | 0 | 5 | 10 | 15 | 20 | 25 | 30 | 35 | 40 | 45 |
| 6 | 0 | 6 | 12 | 18 | 24 | 30 | 36 | 42 | 48 | 54 |
| 7 | 0 | 7 | 14 | 21 | 28 | 35 | 42 | 49 | 56 | 63 |
| 8 | 0 | 8 | 16 | 24 | 32 | 40 | 48 | 56 | 64 | 72 |
| 9 | 0 | 9 | 18 | 27 | 36 | 45 | 54 | 63 | 72 | 81 |

**1.** 20

**2.** 42

Find the row that begins with 6 and the column that begins with 7. The answer, 42, is in the box where the row and column meet.

**3.** 0 **4.** 12 **5.** 35
**6.** 12 **7.** 28 **8.** 21
**9.** 0 **10.** 45 **11.** 16
**12.** 1 **13.** 48 **14.** 9
**15.** 14 **16.** 64 **17.** 6
**18.** 72 **19.** 56 **20.** 2
**21.** 54 **22.** 32 **23.** 9
**24.** 48 **25.** 30 **26.** 21
**27.** 14 **28.** 36

### Page 55

**1.** $96

**2.** $20

Multiply the ones. $2 \times 0 = 0$. Write 0 in the ones place. Multiply the tens. $2 \times 1 = 2$. Write 2 in the tens place. Write a dollar sign in front of the answer.

**3.** 84 **4.** 13
**5.** 153 **6.** 606
**7.** $936 **8.** $3,699
**9.** 4,648 **10.** 6,244
**11.** $86 **12.** 3,633
**13.** $68 **14.** 484
**15.** 6,306 **16.** 48
**17.** 369 **18.** $6,266
**19.** 36 baseballs **20.** 64 children

$$\begin{array}{r} 12 \\ \times\ 3 \\ \hline 36 \end{array} \qquad \begin{array}{r} 32 \\ \times\ 2 \\ \hline 64 \end{array}$$

### Page 56

**1.** $672

**2.** $4,956

Multiply 413 by 2 ones. $2 \times 413 = 826$. Multiply 413 by 1 ten. The second partial product is 4,130. Add the partial products. Write a dollar sign in front of the answer.

**3.** 169 **4.** 3,322
**5.** 1,596 **6.** $299
**7.** 16,828 **8.** $473
**9.** $132 **10.** 3,648
**11.** $693 **12.** $2,940
**13.** $264 **14.** 2,880 customers

$$\begin{array}{r} \$24 \\ \times\ 11 \\ \hline 24 \\ +\ 24 \\ \hline \$264 \end{array} \qquad \begin{array}{r} 120 \\ \times\ 24 \\ \hline 480 \\ +240 \\ \hline 2{,}880 \end{array}$$

**Page 57**

1. 712
2. 98

   Multiply the ones. $7 \times 4 = 28$. Write 8 in the ones place and carry the 2 tens. Multiply the tens. $7 \times 1 = 7$. Add the carried 2. $7 + 2 = 9$ tens. Write 9 in the tens place.

3. 6,420
4. $748
5. 728
6. $6,860
7. 8,145
8. 166
9. 3,682
10. 3,945
11. 60
12. 12,024
13. $4,140
14. 14,910
15. 650 cartons of milk

$$\begin{array}{r} 130 \\ \times \quad 5 \\ \hline 650 \end{array}$$

16. 870 pounds

$$\begin{array}{r} {\scriptstyle 2\,3} \\ 145 \\ \times \quad 6 \\ \hline 870 \end{array}$$

**Page 58**

1. 1,081
2. 66,822

   Multiply 1,806 by 7 ones. The first partial product is 12,642. Multiply 1,806 by 3 tens. The second partial product is 54,180. Add the partial products.

3. $23,576
4. 110,736
5. $32,310
6. $1,464
7. 231,768
8. $3,610
9. $12,642
10. 418,965

    Multiply 8,215 by 1 one.
    $8{,}215 \times 1 = 8{,}215$
    Multiply 8,215 by 5 tens.
    $8{,}215 \times 5 = 41{,}075$
    Add the partial products.

11. 2,240
12. 21,750 tickets
13. 30,000 cans

$$\begin{array}{r} 1{,}450 \\ \times \quad 15 \\ \hline 7\,250 \\ +\,14\,50 \\ \hline 21{,}750 \end{array} \qquad \begin{array}{r} 2{,}500 \\ \times \quad 12 \\ \hline 5\,000 \\ +\,25\,00 \\ \hline 30{,}000 \end{array}$$

**Page 59**

1. 2,440
2. 423,100

   Write two zeros in the answer. Multiply 1 times 4,231. $1 \times 4{,}231 = 4{,}231$. Write these digits to the left of the two zeros.

3. $63,900
4. $73,500
5. 126,840
6. 15,550
7. $112,700
8. 9,270
9. $500
10. 2,010
11. 252,000
12. 165,300
13. 166,600
14. 75,000 sheets
15. 960 bottles

$$\begin{array}{r} 150 \\ \times \quad 500 \\ \hline 75{,}000 \end{array} \qquad \begin{array}{r} 12 \\ \times\ 80 \\ \hline 960 \end{array}$$

**Page 60**

1. 6
2. 8

   Find the row that begins with 9. Read across the row until you find 72. Move up the column to the answer, 8.

**3.** 3 **4.** 5 **5.** 4
**6.** 2 **7.** 2 **8.** 8
**9.** 9 **10.** 7 **11.** 5
**12.** 6 **13.** 3 **14.** 1
**15.** 8 **16.** 7 **17.** 9
**18.** 1 **19.** 5 **20.** 3
**21.** 5 **22.** 3 **23.** 6
**24.** 4 **25.** 4 **26.** 8
**27.** 3

**Page 61**

**1.** 74

**2.** $91

Write a dollar sign in the quotient. Divide 3 into 27. 27 ÷ 3 = 9. Write 9 in the tens place in the quotient. Divide 3 into 3. 3 ÷ 3 = 1. Write 1 in the ones place in the quotient.

**3.** 11 **4.** 411 **5.** 111
**6.** 511 **7.** 81 **8.** $21
**9.** 52 **10.** $71 **11.** 421
**12.** 93

**13.** 11 presents

$$8\overline{)88}\ \ \text{quotient } 11$$

**14.** 64 customers

$$2\overline{)128}\ \ \text{quotient } 64$$

**Page 62**

**1.** 27 R5

**2.** 73 R3

You can't divide 4 into 2. Divide 4 into 29. The closest basic fact is 4 × 7 = 28. Write 7 in the tens place in the quotient. Multiply and subtract, getting a difference of 1.

Bring down the 5. Divide 4 into 15. 15 ÷ 4 is 3 with a remainder of 3. Write 3 in the ones place in the quotient. Write R3 to show the remainder.

**3.** 45 R2 **4.** 17 R1
**5.** 23 R2 **6.** 98 R3
**7.** 28 R5 **8.** 12 R3
**9.** 46 R3 **10.** 12 R3
**11.** 16 R2 **12.** 25 R3

**Page 63**

**1.** 402 R1

**2.** 608

You can't divide 4 into 2. Divide 4 into 24, getting 6. Bring down the 3. You can't divide 4 into 3, so write 0 in the tens place in the quotient. Bring down the 2. Divide 4 into 32 and write the answer, 8, in the ones place in the quotient.

**3.** 513 R5 **4.** 207
**5.** 904 **6.** $850
**7.** 118 **8.** 201

**9.** 150 packages

$$\begin{array}{r} 150 \\ 4\overline{)600} \\ -4\phantom{00} \\ \hline 20\phantom{0} \\ -20\phantom{0} \\ \hline 0\phantom{0} \end{array}$$

**10.** 800 guests

$$\begin{array}{r} 800 \\ 3\overline{)2,400} \\ -2\,4\phantom{00} \\ \hline 00\phantom{0} \\ 0\phantom{0} \\ \hline 00 \\ 0 \end{array}$$

**Page 64**

**1.** 14 R2

**2.** 32 R3

Estimate: 2 goes into 7 about 3 times. Try 3 as the first digit of the quotient. 3 × 24 = 72. Subtract 72 from 77, which leaves 5. Bring down the 1. Divide 24 into 51. 2 × 24 = 48. Write 2 in the quotient. Subtract 48 from 51, which leaves 3. Write R3 for the remainder in the quotient.

**3.** 12 R3 **4.** 46 R6

**5.** 21 R9 **6.** 14 R10

**7.** $37 **8.** 27 R12

**9.** 35 hours **10.** 15 chapters

$$\begin{array}{r} 35 \\ 14\overline{)490} \\ -42 \\ \hline 70 \\ -70 \\ \hline 0 \end{array} \qquad \begin{array}{r} 15 \\ 22\overline{)330} \\ -22 \\ \hline 110 \\ -110 \\ \hline 0 \end{array}$$

**Page 65**

**1.** 9 × 2 = 18

**2.** 4 + 10 = 14

Multiply first. 5 × 2 = 10. Add the 10 to the 4. 10 + 4 = 14

**3.** 6 + 2 = 8 **4.** 3 + 2 = 5

**5.** 10 ÷ 5 = 2 **6.** 20 – 1 = 19

**7.** 8 + 1 = 9 **8.** 3 – 3 = 0

**9.** 6 × 1 = 6 **10.** 12 + 3 = 15

**11.** 21 + 3 = 24 **12.** 30 – 1 = 29

**13.** 2 – 2 = 0 **14.** 16 ÷ 8 = 2

**15.** 7 × 7 = 49 **16.** 45 – 5 = 40

**17.** 12 × 2 = 24 **18.** 30 + 3 = 33

**19.** 14 + 4 = 18 **20.** 33 + 4 = 37

## GED Skill Strategy

**Page 66**

**1.** 400 ÷ 10

**2.** 70 × 10

**3.** 3,000 ÷ 40

**4.** 800 × 90

**5.** 6,000 ÷ 70

**6.** 6,000 × 30

**Page 67**

**7.** 3,600

**8.** 90 × 30 = 2,700

**9.** 70 × 50 = 3,500

**10.** 70 × 30 = 2,100

**11.** 300 × 90 = 2,700

**12.** 800 × 90 = 27,000

**13.** 500 × 40 = 20,000

**14.** 900 × 40 = 36,000

**15.** 400 ÷ 50 = 8

**16.** 400 ÷ 80 = 5

**17.** 200 ÷ 40 = 5

**18.** 2,000 ÷ 50 = 40

**19.** 400 ÷ 80 = 5

**20.** 1,000 ÷ 50 = 20

**21.** 3,000 ÷ 60 = 50

**22.** 3,000 ÷ 30 = 100

**23.** 40 × 300 = 12,000 cars

**24.** 2,000 ÷ 10 = $200

**Page 69**

1. 1,610
2. 1,416
3. 3,038
4. 6,450
5. 17,172
6. 276,597
7. 109,228
8. 740,652
9. 32
10. $18
11. 21
12. 50
13. 243
14. 645
15. 46
16. $84
17. $26
18. 864 ounces
19. $17,520

## GED Test-Taking Strategy

**Page 71**

1. (4) $15
2. (3) 9
3. (2) $105
4. (1) 175
5. (2) $7
6. (4) 30

**Page 72**

1. 804
2. 628
3. 51,600
4. $300 \times 600 = 18{,}000$; 18,411
5. 27 R1
6. $5 - 1 = 4$

**Page 73**

1. 42
2. 35
3. 27
4. 18
5. 4
6. 3
7. 8
8. 9
9. 288
10. 2,048
11. 832
12. 32,643
13. 416
14. 4,555
15. 3,360
16. 207,120
17. 27
18. 23 R1
19. 671 R1
20. 42 R1
21. 16
22. 44
23. 367 R4
24. 245 R3
25. $45 \div 9 = 5$
26. $11 + 48 = 59$
27. $65 - 5 = 60$
28. $6 + 12 = 18$

## GED Test Practice, *pages 74–76*

### Page 74

1. **(4) 126** Multiply the number of packages by the number of pens in a package ($42 \times 3 = 126$). In Choices (1), (2), and (3), incorrect operations are used.
2. **(3) 102** Divide the total number of people by the number of workshops ($306 \div 3 = 102$). Choices (1) and (2) use incorrect operations. Choice (4) includes an error in division.
3. **(1) $43** Divide the total overtime amount by the number of employees ($817 ÷ 19 = $43). Choice (2) uses subtraction. Choice (3) shows the total amount given in the problem. In choice (4), the numbers are multiplied.
4. **(4) 28,800** Multiply the number of boxes by the number of paper clips in a box ($144 \times 200 - 28{,}800$). Choice (1) uses subtraction. Choice (2) uses addition. In Choice (3), incorrect numbers were used to multiply.

### Page 75

5. **(4) 35** Divide the total number of balls by the number in each box ($280 \div 8 = 35$). Choices (1), (2), and (3) use incorrect operations.
6. **(2) $1,320** Multiply the number of sticks by the amount the store makes on each stick (110 × $12 = $1,320). Choice (1) is incorrect because of a multiplication error. Choices (3) and (4) use incorrect operations.
7. **(1) $14,208** Multiply the amount of the payment by the number of payments ($296 × 48 = $14,208). Choice (2) is the amount paid for one year. In Choice (3), the amount paid each month is multiplied by the number of years. Choice (4) uses addition.
8. **(2) 6** Divide the numbers of mouth guards by the number of players ($38 \div 16 = 2R6$). In Choice (1), the numbers are subtracted. In Choice (3), the remainder was incorrectly calculated. Choice (4) shows the number of mouth guards each player received.

### Page 76

9. **(1) 2** Work inside the parentheses first. Multiply $2 \times 9 = 18$ and $3 \times 12 = 36$. Then subtract $38 - 36 = 2$. In Choices (2), (3), and (4), the incorrect order of operations is used.
10. **(1) $2,000** Round numbers to the lead digit (108 rounds to 100, $19 rounds to $20). Then multiply the rounded numbers (100 × $20 = $2,000). In Choices (2), (3), and (4) the numbers are rounded incorrectly.

## Unit 4

### Page 78

| | tens | ones | | tenths | hundredths | thousandths |
|---|---|---|---|---|---|---|
| 1. | | 6 | . | 4 | | |
| 2. | | 1 | . | 7 | 3 | |
| 3. | 1 | 0 | . | 2 | 8 | |
| 4. | | 0 | . | 9 | 5 | 5 |

1. 6 ones 4 tenths
2. 1 one 7 tenths 3 hundredths
   The ones place is just to the left of the decimal point, and the digit 1 is in the ones place. The tenths place is just to the right of the decimal point, and the digit 7 is in the tenths place. The hundredths place is just to the right of the tenths place, and the digit 3 is in the hundredths place.
3. 1 ten 0 ones 2 tenths 8 hundredths
4. 0 ones 9 tenths 5 hundredths 5 thousandths
5. 7
6. 9
   The tenths place is just to the right of the decimal point. The digit 9 is in the tenths place.
7. 0
8. 3
9. 1
10. 4
11. 0
12. 8
13. 1
14. 5
15. 4
16. 2

### Page 79

| | | | | | | | |
|---|---|---|---|---|---|---|---|
| 1. | | 1 | 4 | . | 5 | | |
| 2. | | | 8 | . | 2 | 2 | 5 |
| 3. | | 3 | 0 | . | 2 | | |
| 4. | | | 7 | . | 0 | 4 | |
| 5. | 5 | 0 | 8 | . | 0 | 1 | |
| 6. | | | 0 | . | 6 | 3 | |
| 7. | | | 4 | . | 9 | | |
| 8. | | 4 | 0 | . | 1 | 0 | 8 |

1. fourteen and five tenths
2. eight and two hundred twenty five thousandths
   The right-hand digit is in the thousandths place. The decimal part of the number is 225 thousandths.
3. thirty and two tenths
4. seven and four hundredths
5. five hundred eight and one hundredth
6. sixty-three hundredths
7. four and nine tenths
8. forty and one hundred eight thousandths
9. 70.33
10. 9.052
    The digit 9 for nine ones goes to the left of the decimal point. The right-hand digit must go in the thousandths place, so the digits 052 go the right of the decimal point.
11. 0.26
12. 0.409

## Page 80

1. $\neq$
2. $=$
   Drop the zeros in $12.00. Both numbers have a place value of one ten and two ones.
3. $\neq$
4. $\neq$
5. $=$
6. $=$
7. $\neq$
8. $\neq$
9. is equal to
10. is equal to
    $60.02 = 60.020$
11. is not equal to

## Page 81

1. $=$
2. $\neq$
   The decimals 2.10 and 1.20 are not equal. The number 2.10 has 2 ones, while the number 1.20 has only 1 one.
3. $=$
4. $<$
5. $<$
   Write 0.9 as 0.90. Then compare. $90 < 91$, so $0.9 < 0.91$.
6. $=$
7. $<$
8. $>$
9. $<$
10. $>$
11. $<$
12. 7.01 is 1
    7.1 is 2
    7.11 is 3
13. 0.003 is 1
    0.03 is 2
    0.3 is 3
14. 0.05 is 1
    5.05 is 2
    50.5 is 3

## Page 82

1. 51.5
2. 51.4
   The digit in the hundredths place is 2. Since 2 is less than 5, round down to 4.
3. 51.4
4. 51.5
5. 51.4
6. 51.5
7. 0.4
8. 23.6
   The digit in the hundredths place is 0. Since 0 is less than 5, round down to 6.
9. 3.2
10. 184.0
11. 17.6
12. 8.2

## Page 83

1. 0.09
2. 0.09
   The digit in the thousandths place is 7. Since 7 is greater than 5, round 8 up to 9.
3. 0.09
4. 0.09
5. 0.08
6. 0.09
7. 1.43
8. 0.21
   The digit in the thousandths place is 8. Since 8 is greater than 5, round 0 up to 1.
9. 16.62
10. 50.00
11. 38.37
12. 19.43

## Page 84

**1.** 13.1

**2.** $8.71

Line up the decimal points. Add. Begin with the digits on the right. 3 + 8 = 11. Write 1 in the hundredths place. Carry 1. 7 + 9 = 16. Add the carried 1. 16 + 1 = 17. Write 7 in the tenths place and carry 1 to the dollars column. 7 + 1 = 8. Write 8 in the dollars column.

$$\begin{array}{r} {\scriptstyle 1\,1} \\ \$7.73 \\ +\ \ .98 \\ \hline \$8.71 \end{array}$$

**3.** $2.08

**4.** $44.00

**5.** 60.0

**6.** $1.02

**7.** $352.25

**8.** 4.65

**9.** 29.9

**10.** 557.92

**11.** 6.06

**12.** $56.50

Write $5 with a decimal point and two right-hand zeros: $5.00. Line up the decimal points. Add, starting with the digits on the right.

$$\begin{array}{r} \$51.50 \\ +\ \ 5.00 \\ \hline \$56.50 \end{array}$$

**13.** 109.4

**14.** 66.53

**15.** $26.25

$$\begin{array}{r} \$25.00 \\ +\ \ 1.25 \\ \hline \$26.25 \end{array}$$

**16.** 78.25 hours

$$\begin{array}{r} {\scriptstyle 1} \\ 36.75 \\ +\,41.50 \\ \hline 78.25 \end{array}$$

## Page 85

**1.** 3.28

**2.** $2.07

Line up the decimal points. Subtract. Begin with the digits on the right. Rename. 10 – 3 = 7. 9 – 9 = 0. 4 – 2 = 2. Put a decimal point and a dollar sign in the answer.

$$\begin{array}{r} {\scriptstyle 4\ 9\,10} \\ \$\not{5}.\not{0}\not{0} \\ -\ \ 2.93 \\ \hline \$2.07 \end{array}$$

**3.** 1.1

**4.** 3.89

**5.** $9.09

**6.** 1.18

**7.** 5.25

**8.** 0.04

**9.** 1.09

**10.** 30.76

**11.** $4.69

**12.** 2.4

Write 3 as 3.0. Line up the decimal points. Subtract, starting with the digits on the right. Rename.

10 – 6 = 4.

2 – 0 = 2.

$$\begin{array}{r} {\scriptstyle 2\ 10} \\ \not{3}.\not{0} \\ -\,0.6 \\ \hline 2.4 \end{array}$$

**13.** 1.68

**14.** $4.89

**15.** 9.45 miles

$$\begin{array}{r} {\scriptstyle 1\ 15\ 11\,10} \\ \not{2}\,\not{6}.\not{2}\,\not{0} \\ -\,1\,6.7\,5 \\ \hline 9.4\,5\text{ miles} \end{array}$$

**16.** $10.55

$$\begin{array}{r} {\scriptstyle 1\ 9\ \ 9\,10} \\ \$\not{2}\,\not{0}.\not{0}\,\not{0} \\ -\ \ \ 9.4\,5 \\ \hline \$1\,0.5\,5 \end{array}$$

## Page 86

**1.** 41.6

**2.** $26.40

Multiply. Count the number of decimal places in the problem, 2. Put a decimal point in the answer to show two decimal places. Drop right-hand zeros in answers unless the answer shows dollars and cents.

$$\begin{array}{rll} \$2.64 & \leftarrow & \text{two decimal places} \\ \times \quad 10 & \leftarrow & \text{no decimal places} \\ \hline \$26.40 & \leftarrow & \text{two decimal places} \end{array}$$

**3.** $157.50

**4.** 121.8

**5.** $16.33

**6.** 102

**7.** $2,010.40

**8.** $34.58

**9.** 17.5

**10.** 0.18

Multiply. Count the number of decimal places in the problem, 2. Put a decimal point in the answer to show two decimal places. Write a 0 to the left of the decimal point.

$$\begin{array}{rll} 0.02 & \leftarrow & \text{two decimal places} \\ \times \quad 9 & \leftarrow & \text{no decimal places} \\ \hline 0.18 & \leftarrow & \text{two decimal places} \end{array}$$

**11.** $147.25

**12.** 31.35

**13.** $7.38

$$\begin{array}{r} \$3.69 \\ \times \quad 2 \\ \hline \$7.38 \end{array}$$

**14.** 250 pounds

$$\begin{array}{l} \quad 12.5 \\ \times \ \ 20 \\ \hline 250.0 \text{ pounds} = 250 \text{ pounds} \end{array}$$

## Page 87

**1.** 0.0038

**2.** 0.052

Multiply. Rename. Count the number of decimal places in the problem, 3. Put a decimal point in the answer to show three decimal places.

$$\begin{array}{rll} \overset{1}{0.26} & \leftarrow & \text{two decimal places} \\ \times \ 0.2 & \leftarrow & \text{one decimal places} \\ \hline 0.052 & \leftarrow & \text{three decimal places} \end{array}$$

**3.** 0.35

**4.** $.19

**5.** 0.2355

**6.** $90.00

**7.** 47.25

**8.** 27.473

**9.** 1.62

**10.** 0.522

Multiply. Rename. Count the number of decimal places in the problem, 3. Put a decimal point in the answer to show three decimal places.

$$\begin{array}{rll} 0.58 & \leftarrow & \text{two decimal places} \\ \times \ 0.9 & \leftarrow & \text{one decimal places} \\ \hline 0.522 & \leftarrow & \text{three decimal places} \end{array}$$

**11.** $24.58

**12.** 8.892

**13.** $109.20

$$\begin{array}{l} \quad \$16.80 \\ \times \quad\ \ 6.5 \\ \hline \$109.200 = \$109.20 \end{array}$$

**14.** 2.7 miles

$$\begin{array}{l} \quad 0.75 \\ \times \ 3.6 \\ \hline 2.700 \text{ miles} = 2.7 \text{ miles} \end{array}$$

## Page 88

**1.** 0.84

**2.** 0.16

Set up the problem. Put a decimal point in the answer. Divide 15 into 24. 24 ÷ 15 = 1 with a remainder. Multiply and subtract. Add an extra zero to the dividend. Bring down this zero. Divide again. 90 ÷ 15 = 6

$$\begin{array}{r} 0.16 \\ 15\overline{)2.40} \\ -1\,5 \\ \hline 90 \\ -90 \\ \hline 0 \end{array}$$

**3.** $1.50

**4.** 0.0625

**5.** 0.63

**6.** 0.306

**7.** 120.3

**8.** 1.87

**9.** 4.09

**10.** 0.08

**11.** 7.46

**12.** 0.7

**13.** 22.82 feet

$$\begin{array}{r} 22.82 \\ 3\overline{)68.46} \end{array}$$

**14.** $1.59

$$\begin{array}{r} \$1.59 \\ 4\overline{)\$6.36} \end{array}$$

## Page 89

**1.** 4.28

**2.** 9.4

Set up the problem. Put a decimal point in the answer. Divide to two decimal places. Round 9.43 to one decimal place.

$$\begin{array}{r} 9.43 \\ 5\overline{)47.16} \\ -45 \\ \hline 21 \\ -20 \\ \hline 16 \\ -15 \\ \hline 1 \end{array}$$

**3.** 2.3

**4.** 3.5

**5.** 0.07

**6.** $.05

Set up the problem. Put a dollar sign and decimal point in the answer. Divide to three decimal places. Round $.052 to two decimal places.

$$\begin{array}{r} \$.052 \\ 50\overline{)\$2.600} \\ -2\,50 \\ \hline 100 \\ -100 \\ \hline 0 \end{array}$$

**7.** 0.14

**8.** $.09

**9.** $1.99

$$\begin{array}{r} \$1.99 \\ 24\overline{)\$47.70} \end{array}$$

**10.** $11.42

$$\begin{array}{r} \$11.42 \\ 3\overline{)\$34.25} \end{array}$$

**Page 90**

1. 53
2. $18

   Set up the problem. Move the decimal point one place to the right. Put a dollar sign and decimal point in the answer. Divide.

   $$0.3\overline{)\$5.40} \rightarrow \begin{array}{r} \$18 \\ 3\overline{)\$54} \\ \underline{-3} \\ 24 \\ \underline{-24} \\ 0 \end{array}$$

3. 0.8
4. 0.4
5. 6.18
6. $5.06

   Set up the problem. Move the decimal point one place to the right. Put a dollar sign and decimal point in the answer. Divide to three decimal places. Round to two decimal places.

   $$1.3\overline{)\$6.58} \rightarrow \begin{array}{r} \$5.061 \\ 13\overline{)\$65.800} \\ \underline{-65} \\ 80 \\ \underline{-78} \\ 20 \\ \underline{-13} \\ 7 \end{array}$$

7. $1.20
8. 1.62
9. $1.80

   $$2.5\overline{)\$4.500} = \begin{array}{r} \$18 \\ 25\overline{)\$45.00} \end{array}$$

10. $18.00

    $$7.5\overline{)\$135.0} = \begin{array}{r} \$18 \\ 75\overline{)\$1350} \end{array}$$

**Page 91**

Answers will vary, depending on how you round the numbers in the problems. Sample answers are given.

1. $6.00 \div 0.03 = 200$
2. $15 \div 3 = 5$
3. $8.3 \div 0.9 = 9.2$
4. $1.2 \div 30 = 0.04$
5. $0.5 \div 2 = 0.25$
6. $0.02 \div 2 = 0.01$
7. $3.6 \div 12 = 0.3$
8. $0.5 \div 60 = 0.008$
9. $2.0 \div 4 = 0.5$
10. $0.36 \div 0.9 = 0.4$
11. $0.05 \div 2 = 0.025$
12. $3 \div 15 = 0.2$
13. $14.00

    Round 4.7 to 5. Round 71.80 to 70. Divide $70 by 5. $\$70 \div 5 = \$14.00$
14. 1 pound per day

    Round both 32.5 and 31 to 30. Divide 30 by 30. 30 pounds ÷ 30 days = 1 pound per day

## GED Skill Strategy

**Page 93**

1. 4.15
2. $3.67

   Clear the calculator. Enter 3.80. Press the × key. Enter 0.97. Press the = key.
3. $38.78
4. 19.28
5. 0.34
6. $1.22
7. 9.36
8. $2.89
9. 38.22
10. 1.08
11. 0.05
12. $.30
13. $1.08
14. $57.31
15. $86.49
16. 6.34

**17.** 27.5 miles per hour

$$1.75\overline{)48.125} = 27.5$$

**18.** $223.65 per week

$$\begin{array}{r} \$31.95 \\ \times \quad 7 \\ \hline \$223.65 \end{array}$$

**19.** $96.29 per night

$$\begin{array}{r} \$89.99 \\ + \ 6.30 \\ \hline \$96.29 \end{array}$$

**20.** $0.95

$$\begin{array}{r} \$12.55 \\ - \ 11.60 \\ \hline \$\ 0.95 \end{array}$$

**Page 94**

1. 60%
2. 7%
3. 135%
4. 75%
5. 90%
6. 120%
7. 1%
8. 25%
9. 10%
10. 45%
11. 58%
12. 200%
13. 80%
14. 75%
15. 16%
16. 150%
17. 8.5%

**Page 95**

1. 0.5
2. 3.8

   Write the number without the percent sign (380). Move the decimal two places to the left (3.80).
3. 0.05
4. .732
5. 1.12
6. 0.18
7. 0.495
8. 1.3
9. 0.74
10. 0.071
11. 3.45
12. 0.935
13. 4.5%
14. 230%
15. 85%
16. 20%
17. 50.6%
18. 7%
19. 55%
20. 185%
21. 31.8%
22. 4%
23. 341%
24. 16%
25. 0.15%
26. 8%

**Page 96**

1. whole = 40
   percent = 20%
   part = 8
2. whole = 48
   percent = 75%
   part = 36
3. whole = 150
   percent = 30%
   part = 45
4. whole = 25
   percent = 60%
   part = 15
5. whole = 200
   percent = 25%
   part = 50
6. whole = 80
   percent = 150%
   part = 120
7. 5
8. 320
9. 180
10. 42
11. 37.5
12. 660
13. $240
14. 36 blue cars

**Page 97**

1. part = 5
   whole = 20
   percent = 5 ÷ 20
2. part = 15
   whole = 40
   percent = 15 ÷ 40
3. part = 8
   whole = 80
   percent = 8 ÷ 80
4. part = 4
   whole = 100
   percent = 4 ÷ 100
5. part = 56
   whole = 28
   percent = 56 ÷ 28
6. part = 270
   whole = 360
   percent = 270 ÷ 360
7. 50%
8. 175%
9. 300%
10. 25%
11. 75%
12. 60%

## GED Skill Strategy

**Page 98**

1. **a.** 45%
   **b.** 35%
2. **a.** 15%
   **b.** 40%
3. **a.** 34
   **b.** 30%
4. **a.** 85
   **b.** 7%

**Page 99**

5. **a.** 40
   **b.** 25%
   **c.** 55%
   **d.** 75%
6. **a.** 60
   **b.** 85%
   **c.** 5%
   **d.** 75%

7. **a.** $800
   **b.** 25%
   **c.** 55%
   **d.** 35%
   **e.** 20%
8. **a.** $600
   **b.** 13%
   **c.** 15%
   **d.** 60%
   **e.** 3%

## GED Skill Strategy

### Page 100

1. 21
2. 11
3. 1.5
4. 90
5. 90
6. 208
7. 96
8. 32
9. 144

### Page 101

10. 18.5
11. 81.3
12. 5.4
13. 1.9
14. 108.5
15. 9.7
16. 13.8
17. 3.4
18. 10.5
19. $120
20. $127,500
21. 7,680 people
22. 36 questions

## GED Skill Strategy

### Page 102

1. $4.80
2. $2.45
3. $3.00
4. $8.75

### Page 103

5. Florida
6. Colorado
7. $20
8. Ohio, Connecticut, New Jersey
9. $7.50
10. $1.48

## GED Test-Taking Strategy

### Page 105

1. **(1) 0.32 minute** Choice (1) is correct because subtraction is used to find how much longer Song 1 is than Song 2 (2.75 −2.43 = 0.32). Choices (2), (3), and (4) use incorrect operations.
2. **(3) $12.80** Choice (3) is correct because division is used to find the hourly rate ($460.00 ÷ 36 = $12.80). Choices (1) and (4) use incorrect operations. Choice (2) contains an error in division.
3. **(2) 12.5** Choice (2) is correct because subtraction is used to find the feet of canvas left over (16.5 −4 = 12.5). Choices (1), (3), and (4) use incorrect operations.
4. **(2) $1.81** Choice (2) is correct because multiplication is used to find the cost of the fruit salad (0.91 × $1.99 = $1.81). Choices (1) and (3) use incorrect operations. Choice (4) has an error in addition.

5. **(4) 22.5** Choice (4) is correct because multiplication is used to find the total amount ($7.5 \times 3 = 22.5$). Choices (1), (2), and (3) use incorrect operations.
6. **(1) 4.7** Choice (1) is correct because subtraction is used to find the difference in weights ($175.2 - 170.5 = 4.7$). Choices (2), (3), and (4) contain errors in subtraction.

**Page 106**

1. $0.071 < 0.17$ or $0.17 > 0.071$
2. 2.25
3. 11.43
4. 0.544
5. 6.12
6. 12.0

**Page 107**

1. $<$
2. $=$
3. $>$
4. $>$
5. 2.35
6. 0.07
7. 12.01
8. $1.60
9. $.40
10. $8.10
11. 2.41
12. 4.79
13. 0.02
14. 6.56
15. 14.42
16. 21.6
17. 0.06
18. 80.94
19. 2.28
20. 0.41
21. 0.02
22. 0.31
23. $140
24. 60%

## GED Test Practice, *pages 108–110*

**Page 108**

1. **(2) 24.8** Choice (2) is correct. Choice (1) is two hundred forty eight seconds. Choice (3) is twenty four and eight hundredths seconds, and Choice (4) is two and forty eight hundredths seconds.
2. **(4) chicken, sirloin, pork** Choice (4) is correct because chicken has the least fat (3.6g), sirloin is next (6.1g), and pork has the greatest amount (7.2). Choices (1), (2), and (3) list the meats in incorrect orders.
3. **(1) $1,64** Choice (1) is correct because subtraction was used to find the amount of change ($50 – 48,36 = $1.64). Choices (2) and (3), have errors in subtraction. Choice (4) used addition.
4. **(3) 6.3** Choice (3) is correct because addition was used to find the number of miles walked ($3.5 + 2.8 = 6.3$). Choice (1) used subtraction, and incorrect operation. Choices (2) and (4) have errors in addition.

**Page 109**

5. **(4) 62.75 inches** Choice (4) is correct because division was used to find the length of the pieces ($188.25 \div 3 = 62.75$). Choices (1), (2), and (3) used incorrect operations.
6. **(2) 60.75** Choice (2) is correct because multiplication was used to find the total number of ounces ($6.75 \times 9 = 60.75$). Choices (1) and (3) contained errors in multiplication. Choice (4) used the incorrect operation.

7. **(3) about 30** Choice (3) is correct because estimation (283.5 rounds to 300 and 11 rounds to 10) and division ($300 \div 10 = 30$) were used to find the miles per gallon.

8. **(2) 49** Choice (2) is correct because multiplication was used to find the number of employees ($140 \times 35\% = 49$). Choice (1) is the percent for the problem. Choice (3) contains an error in calculation. Choice (4) used the incorrect operation.

## Page 110

9. **(1) 6%** Choice (1) is correct because division was used to find the percentage ($3 \div 50 = 0.06$) The decimal was then moved to the right two places and the percent sign added (6%). Choice (2) has an error in calculation. Choice (3) is incorrect because the wrong operation was used. In choice (4), the decimal was moved incorrectly.

10. **(3) 80%** Choice (3) is correct because addition was used to find out what percentage labor plus materials is.

# Unit 5

## Page 112

1. $\frac{4}{5}$
2. $\frac{2}{3}$

   There are 3 equal parts, so the denominator is 3. 2 parts are shaded, so the numerator is 2.
3. $\frac{1}{10}$
4. $\frac{4}{7}$
5. Shade 3 parts of the figure.
6. Shade 5 parts of the figure.
7. Shade 1 of the 2 figures.
8. Shade 3 of the 10 figures.

## Page 113

1. $\frac{3}{4} = \frac{6}{8}$
2. $\frac{1}{2} = \frac{3}{6}$

   The first figure has 1 out of 2 parts shaded. The second figure has 3 out of 6 parts shaded.
3. $\frac{1}{3} = \frac{5}{15}$
4. $\frac{1}{4} = \frac{2}{8}$
5. $\frac{1}{2} = \frac{8}{16}$
6. $\frac{2}{4} = \frac{4}{8}$

## Page 114

1. $\frac{3}{7}$
2. $\frac{2}{3}$

   Dividing both parts of the fraction by 4 will reduce it to lowest terms.

   $\frac{8}{12} = \frac{8 \div 4}{12 \div 4} = \frac{2}{3}$
3. $\frac{5}{9}$
4. $\frac{2}{5}$
5. $\frac{3}{4}$
6. $\frac{1}{3}$
7. $\frac{4}{5}$
8. $\frac{5}{7}$
9. 3
10. 7

    $\frac{7}{14} = \frac{7 \div 7}{14 \div 7} = \frac{1}{2}$
11. 3
12. 6
13. 4
14. 9

## Page 115

**1.** $\frac{4}{6}$

**2.** $\frac{6}{10}$

Multiply the numerator by 2. $2 \times 3 = 6$.
Multiply the denominator by 2. $2 \times 5 = 10$.
The new fraction is $\frac{6}{10}$.

$\frac{3}{5} = \frac{3 \times 2}{5 \times 2} = \frac{6}{10}$

**3.** $\frac{8}{14}$

**4.** $\frac{10}{16}$

**5.** 2

**6.** 4

$\frac{4}{5} = \frac{4 \times 4}{5 \times 4} = \frac{16}{20}$

**7.** 3

**8.** 5

**9.** 6

**10.** 6

$\frac{3}{8} = \frac{3 \times 2}{8 \times 2} = \frac{6}{16}$

**11.** 10

**12.** 21

**13.** 8

**14.** 27

**15.** 15

**16.** 100

## Page 116

**1.** $>$

**2.** $<$

Multiply the denominator. $7 \times 3 = 21$. Use 21 as the common denominator. Raise each fraction to higher terms, with 21 as the denominator. Compare the numerators. $12 < 14$, so $\frac{4}{7}$ is less than $\frac{2}{3}$.

$\frac{4}{7} = \frac{4 \times 3}{7 \times 3} = \frac{12}{21}$

$\frac{2}{3} = \frac{2 \times 7}{3 \times 7} = \frac{14}{21}$

**3.** $=$

**4.** $>$

**5.** $<$

**6.** $>$

**7.** $>$

**8.** $<$

**9.** Enrique; $\frac{2}{3} > \frac{1}{2}$

**10.** onion; $\frac{3}{8} > \frac{1}{3}$

## Page 117

**1.** $<$

**2.** $>$

List the multiples of each denominator. 15 is the LCD. Raise $\frac{2}{5}$ to higher terms with 15 as the LCD. Compare the numerators.

$\frac{8}{15}$: $\boxed{15}$ 30 45

$\frac{2}{5}$: 5 10 $\boxed{15}$ 20 25

$\frac{2}{5} = \frac{6}{15}$

$\frac{8}{15} > \frac{6}{15}$ so $\frac{8}{15} > \frac{2}{5}$

**3.** $>$

**4.** $<$

**5.** $<$

**6.** $>$

**7.** $>$

**8.** $<$

## Page 118

**1.** M

**2.** P

The fraction is proper because the numerator is less than the denominator.

**3.** I

**4.** W

**5.** I

**6.** $2\frac{3}{4}$

**7.** 3

Divide 30 by 10. $30 \div 10 = 3$ The improper fraction $\frac{30}{10}$ equals the whole number 3.

**8.** $2\frac{2}{3}$

**9.** $4\frac{1}{5}$

**10.** $2\frac{2}{7}$

**11.** $1\frac{7}{8}$

**12.** $4\frac{1}{2}$

**13.** $2\frac{1}{6}$

**14.** 5

**15.** $3\frac{1}{8}$

## Page 119

1. $\frac{11}{2}$
2. $\frac{26}{3}$
3. $\frac{36}{6}$
4. $\frac{51}{4}$
5. $\frac{13}{8}$
6. $\frac{18}{5}$
7. $\frac{15}{2}$
8. $\frac{37}{7}$
9. $\frac{32}{3}$
10. $\frac{35}{6}$
11. $\frac{400}{20}$
12. $\frac{34}{3}$
13. $\frac{52}{5}$
14. $\frac{67}{2}$
15. $\frac{10000}{100}$
16. $\frac{9}{4}$
17. $\frac{23}{4}$

## Page 120

1. $1\frac{2}{5}$
2. $1\frac{1}{10}$
   Add the fraction parts by adding only the numerators. $\frac{6}{10} + \frac{5}{10} = \frac{11}{10}$. Reduce to lowest terms. $\frac{11}{10} = 1\frac{1}{10}$.
3. $\frac{2}{3}$
4. $2\frac{3}{5}$
5. $3\frac{2}{3}$
6. $\frac{5}{6}$
7. $\frac{11}{12}$
8. $\frac{5}{6}$
9. $3\frac{1}{2}$
10. $\frac{4}{5}$ mile
    Add $\frac{2}{10} + \frac{6}{10}$.
    $\frac{2}{10} + \frac{6}{10} = \frac{8}{10}$.
    Reduce $\frac{8}{10}$ to $\frac{4}{5}$.
11. 1 cup
    Add $\frac{1}{3} + \frac{2}{3} = \frac{3}{3}$.
    Reduce $\frac{3}{3}$ to 1.

## Page 121

1. $10\frac{1}{10}$
2. $5\frac{1}{3}$
3. $25\frac{2}{5}$
4. $8\frac{2}{5}$
5. 10
6. $5\frac{1}{2}$
7. $9\frac{1}{2}$
8. 11
9. $12\frac{1}{2}$
10. $16\frac{1}{4}$
11. $3\frac{1}{2}$ pounds
12. $7\frac{3}{4}$ yards

## Page 122

1. $\frac{1}{2}$
2. $\frac{5}{6}$
   Use 12 as a common denominator.
   Raise $\frac{1}{3}$ to higher terms with 12 as the denominator. Add the numerators. Reduce.
   $\frac{1}{3} = \frac{4}{12}$
   $\frac{4}{12} + \frac{6}{12} = \frac{10}{12} = \frac{5}{6}$
3. $1\frac{5}{16}$
4. $\frac{15}{28}$
5. $1\frac{3}{8}$

6. $\frac{7}{10}$
7. $1\frac{5}{24}$
8. $1\frac{1}{12}$
9. $\frac{15}{16}$ inch
10. $\frac{5}{12}$

**Page 123**

1. $8\frac{11}{20}$
2. $16\frac{13}{15}$
Use 15 as a common denominator. Raise $\frac{2}{3}$ and $\frac{1}{5}$ to higher terms with 15 as the denominator. Add the numerators. Add the whole numbers.
$$\begin{array}{r} 9\frac{2}{3} = 9\frac{10}{15} \\ +\,7\frac{1}{5} = 7\frac{3}{15} \\ \hline 16\frac{13}{15} \end{array}$$
3. $9\frac{37}{40}$
4. $11\frac{1}{36}$
5. $7\frac{11}{48}$
6. $10\frac{7}{12}$
7. $12\frac{3}{4}$
8. $13\frac{3}{10}$
9. $3\frac{4}{5}$ miles
Add $2\frac{1}{2}$ and $1\frac{3}{10}$.
$2\frac{1}{2} = 2\frac{5}{10}$
$2\frac{5}{10} + 1\frac{3}{10} = 3\frac{8}{10}$
$3\frac{8}{10} = 3\frac{4}{5}$
10. $9\frac{3}{4}$ pounds
Add $6\frac{1}{4}$ and $3\frac{1}{2}$.
$3\frac{1}{2} = 3\frac{2}{4}$
$6\frac{1}{4} + 3\frac{2}{4} = 9\frac{3}{4}$

**Page 124**

1. $\frac{1}{2}$
2. $\frac{2}{3}$
Subtract the fraction parts by subtracting only the numerators. $\frac{5}{6} - \frac{1}{6} = \frac{4}{6}$. Reduce the fraction part to lowest terms.
$\frac{4}{6} = \frac{2}{3}$
$\frac{5}{6} - \frac{1}{6} = \frac{4}{6} = \frac{2}{3}$
3. $\frac{1}{2}$
4. $\frac{3}{5}$
5. $\frac{1}{3}$
6. $\frac{3}{7}$
7. $\frac{2}{9}$
8. $\frac{2}{5}$
9. $\frac{1}{2}$
10. $\frac{1}{2}$
$\frac{3}{4} - \frac{1}{4} = \frac{2}{4}$
Reduce $\frac{2}{4}$ to $\frac{1}{2}$.
11. $\frac{2}{5}$
$\frac{8}{10} - \frac{4}{10} = \frac{4}{10}$
Reduce $\frac{4}{10}$ to $\frac{2}{5}$.

**Page 125**

1. $1\frac{1}{5}$
2. $2\frac{2}{7}$
3. $5\frac{3}{5}$
4. $\frac{3}{8}$
5. $9\frac{2}{3}$
6. $2\frac{2}{5}$
7. $5\frac{1}{5}$
8. $5\frac{1}{3}$
9. $5\frac{1}{2}$
10. $\frac{1}{3}$
11. $1\frac{1}{2}$ cups
12. $17\frac{1}{4}$ inches

**Page 126**

1. $\frac{1}{12}$
2. $\frac{4}{5}$

   Multiply the denominators. $6 \times 5 = 30$. Use 30 as a common denominator. Raise both fractions to higher terms with 30 as the denominator. Subtract the numerators. Reduce the answer to lowest terms.

   $\frac{6}{6} = \frac{6 \times 5}{6 \times 5} = \frac{30}{30}$

   $\frac{1}{5} = \frac{1 \times 6}{5 \times 6} = \frac{6}{30}$

   $\frac{30}{30} - \frac{6}{30} = \frac{24}{30}$

   $\frac{24}{30} = \frac{24 \div 6}{30 \div 6} = \frac{4}{5}$
3. $\frac{1}{14}$
4. $\frac{7}{24}$
5. $\frac{13}{36}$
6. $\frac{13}{24}$
7. $\frac{3}{10}$
8. $\frac{7}{16}$
9. $\frac{7}{20}$
10. $\frac{4}{21}$
11. $\frac{3}{8}$ pound

    Subtract $\frac{1}{8}$ from $\frac{1}{2}$.

    $\frac{1}{2} = \frac{4}{8}$

    $\frac{4}{8} - \frac{1}{8} = \frac{3}{8}$
12. $\frac{1}{4}$ of day

    Subtract $\frac{1}{12}$ from $\frac{1}{3}$.

    $\frac{1}{3} = \frac{4}{12}$

    $\frac{4}{12} - \frac{1}{12} = \frac{3}{12} = \frac{1}{4}$

**Page 127**

1. $4\frac{13}{18}$
2. $2\frac{4}{15}$

   Find the LCD. Raise each fraction to higher terms with 15 as the denominator.

   $$\begin{array}{rcr} 11\frac{3}{5} & = & 11\frac{9}{15} \\ -\ 9\frac{1}{3} & = & 9\frac{5}{15} \\ \hline & & 2\frac{4}{15} \end{array}$$
3. $3\frac{19}{30}$
4. $2\frac{9}{40}$
5. $2\frac{17}{80}$
6. $2\frac{7}{40}$
7. $5\frac{4}{9}$
8. $2\frac{1}{40}$
9. $4\frac{1}{16}$
10. $1\frac{19}{60}$
11. $55\frac{3}{8}$ inches

    Find the LCD. Raise each fraction to higher terms with 8 as the denominator.

    $$\begin{array}{rcr} 65\frac{7}{8} & = & 65\frac{7}{8} \\ -\ 10\frac{1}{4} & = & 10\frac{4}{8} \\ \hline & & 55\frac{3}{8} \end{array}$$
12. $20\frac{1}{8}$ yards

    Find the LCD. Raise each fraction to higher terms with 8 as the denominator.

    $$\begin{array}{rcr} 32\frac{5}{8} & = & 32\frac{5}{8} \\ -\ 12\frac{1}{4} & = & 12\frac{4}{8} \\ \hline & & 20\frac{1}{8} \end{array}$$

## GED Skill Strategy

### Page 129

1. Year 4
2. Year 1
3. 50 more employees
4. 80 campsites
5. week 1
6. 246 reservations

### Page 130

**1.** $\frac{1}{4}$ **2.** $\frac{5}{24}$

Multiply the numerators. $5 \times 1 = 5$.

Multiply the denominators. $8 \times 3 = 24$.

$\frac{5}{8} \times \frac{1}{3} = \frac{5 \times 1}{8 \times 3} = \frac{5}{24}$

**3.** $\frac{3}{8}$ **4.** $\frac{7}{30}$

**5.** $\frac{3}{80}$ **6.** $\frac{7}{12}$

**7.** $\frac{1}{12}$ **8.** $\frac{5}{21}$

**9.** $\frac{9}{100}$ **10.** $\frac{5}{24}$

**11.** $\frac{1}{6}$ **12.** $\frac{3}{16}$

**13.** $\frac{1}{6}$ pound

To find $\frac{1}{3}$ of $\frac{1}{2}$ multiply. $\frac{1}{3} \times \frac{1}{2} = \frac{1}{6}$

**14.** $\frac{3}{16}$ of the class is married women

To find $\frac{1}{2}$ of $\frac{3}{8}$ multiply. $\frac{3}{8} \times \frac{1}{2} = \frac{3}{16}$

### Page 131

**1.** $6\frac{1}{2}$ **2.** $7\frac{1}{2}$

Change $2\frac{1}{2}$ to an improper fraction.

$2\frac{1}{2} = \frac{2 \times 2 + 1}{2} = \frac{5}{2}$. Change 3 to an improper fraction. $3 = \frac{3}{1}$.

Multiply the numerators. $5 \times 3 = 15$.

Multiply the denominators. $2 \times 1 = 2$.

Change the answer to an improper fraction.

$2\frac{1}{2} \times 3 = \frac{5}{2} \times \frac{3}{1} = \frac{5 \times 3}{2 \times 1} = \frac{15}{2} = 7\frac{1}{2}$

**3.** $5\frac{3}{5}$ **4.** $4\frac{1}{5}$

**5.** $14\frac{7}{8}$ **6.** $3\frac{4}{15}$

**7.** 8 **8.** $9\frac{11}{12}$

**9.** $65\frac{1}{3}$ **10.** 44

**11.** $4\frac{1}{3}$ **12.** $6\frac{13}{20}$

**13.** $3\frac{1}{2}$ miles

Change 7 to an improper fraction, then multiply $\frac{1}{2}$ by 7.

$\frac{1}{2} \times 7 = \frac{1}{2} \times \frac{7}{1} = \frac{7}{2} = 3\frac{1}{2}$.

**14.** 21 plants

Change $3\frac{1}{2}$ and 6 to improper fractions, then multiply.

$3\frac{1}{2} \times 6 = \frac{7}{2} \times \frac{6}{1} = \frac{42}{2} = 21$.

### Page 132

**1.** $\frac{4}{7}$

**2.** $\frac{8}{9}$

Invert the fraction to the right of the divide sign and multiply.

$\frac{2}{3} \div \frac{3}{4} = \frac{2}{3} \times \frac{4}{3} = \frac{2 \times 4}{3 \times 3} = \frac{8}{9}$

**3.** $2\frac{1}{4}$ **4.** $\frac{7}{8}$

**5.** $\frac{8}{15}$ **6.** $2\frac{1}{12}$

**7.** $\frac{4}{7}$ **8.** $1\frac{1}{15}$

**9.** $\frac{12}{13}$ **10.** $1\frac{1}{2}$

**11.** $\frac{8}{9}$ **12.** $1\frac{5}{21}$

**13.** 3 pieces

Divide $\frac{3}{4}$ by $\frac{1}{4}$. Invert the fraction $\frac{1}{4}$ and multiply.

$\frac{3}{4} \div \frac{1}{4} = \frac{3}{4} \times \frac{4}{1} = \frac{12}{4} = 3$

**14.** 6 pieces

Divide $\frac{3}{4}$ by $\frac{1}{8}$. Invert the fraction $\frac{1}{8}$ and multiply.

$\frac{3}{4} \div \frac{1}{8} = \frac{3}{4} \times \frac{8}{1} = \frac{24}{4} = 6$

**Page 133**

**1.** 2

**2.** 2

Change both mixed numbers to improper fractions. Invert the fraction to the right of the divide sign and multiply. Change the answer to a whole number.

$2\frac{2}{3} \div 1\frac{1}{3} = \frac{8}{3} \div \frac{4}{3} =$

$\frac{8}{3} \times \frac{3}{4} = \frac{8 \times 3}{3 \times 4} = \frac{24}{12} = 2$

**3.** $3\frac{1}{8}$ **4.** 4

**5.** $1\frac{1}{7}$ **6.** $8\frac{2}{3}$

**7.** $1\frac{5}{11}$ **8.** $6\frac{2}{3}$

**9.** 20 packages

Change $2\frac{1}{2}$ to an improper fraction and divide by $\frac{1}{8}$.

$2\frac{1}{2} = \frac{2 \times 2 + 1}{2} = \frac{5}{2}$

$\frac{5}{2} \div \frac{1}{8} = \frac{5}{2} \times \frac{8}{1} = \frac{40}{2} = 20$

**10.** 50 pieces

Change 25 to an improper fraction and divide by $\frac{1}{2}$.

$25 = \frac{25}{1}$

$\frac{25}{1} \div \frac{1}{2} = \frac{25}{1} \times \frac{2}{1} = \frac{50}{1} = 50$

**Page 134**

**1.** $\frac{7}{20}$

**2.** 5

Write 500 over 100. $\frac{500}{100} = \frac{5}{1} = 5$

**3.** $\frac{1}{25}$ **4.** $\frac{11}{20}$

**5.** $2\frac{1}{2}$ **6.** $\frac{7}{10}$

**7.** $\frac{3}{10}$ **8.** $2\frac{3}{5}$

**9.** $66\frac{2}{3}\%$

**10.** 80%

Divide 4 by 5. Write the answer with two decimal places. Move the decimal point two places to the right. Add a percent sign.

$4 \div 5 = 0.80$

$0.80 = 80\%$

**11.** 45%

**12.** $83\frac{1}{3}\%$

**13.** $33\frac{1}{3}\%$

$1 \div 3 = 0.33\frac{1}{3}$

$0.33\frac{1}{3} = 33\frac{1}{3}\%$

**14.** $1\frac{3}{10}$

$130\% = \frac{130}{100}$

$\frac{130}{100} = 1.3$

$1.3 = 1\frac{3}{10}$

## Page 135

1. $\frac{2}{5}$
2. $\frac{9}{10}$

   Write the first number, 9, as the numerator. Write the second number, 10, as the denominator.
3. $\frac{4}{7}$
4. $\frac{3}{1}$
5. $\frac{56}{1}$
6. $\frac{28}{1}$
7. $\frac{3}{4}$
8. $\frac{1}{3}$

   Divide both the numerator and the denominator by 3.

   $\frac{3}{9} = \frac{3 \div 3}{9 \div 3} = \frac{1}{3}$
9. $\frac{5}{2}$
10. $\frac{4}{5}$
11. $\frac{5}{1}$
12. $\frac{8}{9}$

    Write 64, the number of people who attended, as the numerator. Write 72, the number of people who were invited, as the denominator. Reduce.

    $\frac{64}{72} = \frac{64 \div 8}{72 \div 8} = \frac{8}{9}$
13. $\frac{4}{3}$

    Write 56, the number of newspapers delivered on Sunday, as the numerator. Write 42, the number of newspapers delivered on weekdays, as the denominator. Reduce.

    $\frac{56}{42} = \frac{56 \div 14}{42 \div 14} = \frac{4}{3}$

## Page 136

1. 2
2. 10

   Look at the denominators. 12 has been multiplied by 2 to get 24. Multiply 5 by 2.

   $5 \times 2 = 10$

   $\frac{5}{12} = \frac{5 \times 2}{12 \times 2} = \frac{10}{24}$
3. 3
4. 20
5. 32
6. 30
7. 18
8. 24
9. not equal
10. equal

    $4 \times 30 = 120$

    $6 \times 20 = 120$

    $4 \times 30 = 6 \times 20$, so the ratios are equal.
11. equal
12. not equal

## Page 137

1. 15
2. 6

   $12 \times n = 9 \times 8$

   $12n = 72$

   $n = 72 \div 12 = 6$

   $n = 6$
3. 3
4. 30
5. 20
6. 8
7. 12
8. 10
9. 50 minutes

   $\frac{3}{10} = \frac{15}{n}$

   $3n = 150$

   $n = 150 \div 3 = 50$

   $n = 50$ minutes

10. $0.64

$$\frac{3}{\$0.96} = \frac{2}{n}$$
$$3n = \$1.92$$
$$n = \$1.92 \div 3 = \$0.64$$
$$n = \$0.64$$

## GED Test-Taking Strategy

### Page 139

1. **(3) 0.875** Choices (1), (2), and (4) are incorrect because the decimal points are in the incorrect places.
2. **(2) $\frac{1}{4}$** Choices (2), (3), and (4) are incorrect because 25/100 reduces to 1/4.
3. **(2) $\frac{89}{100}$** Choices (1), (3), and (4) are incorrect because 100 must be placed in the denominator to show a fraction of a dollar.
4. **(3) $4\frac{3}{4}$** Choices (1), (2), and (4) are incorrect because the numbers do not represent 4.75 as a fraction.
5. **(1) 6%** Choices (2), (3), and (4) are incorrect because 3 divided by 5 equals 0.6, or 6%.
6. **(1) 0.7** Choices (2), (3), and (4) are incorrect because the decimal points are in the incorrect places.

### Page 140

1. $8\frac{11}{30}$
2. $\frac{1}{2}$
3. $\frac{6}{25}$
4. $1\frac{3}{16}$
5. 25%
6. $n = 12$

### Page 141

1. $\frac{13}{6}$
2. $\frac{31}{8}$
3. $\frac{16}{3}$
4. $\frac{8}{5}$
5. $\frac{2}{5}$
6. $\frac{5}{12}$
7. $\frac{21}{40}$
8. $1\frac{3}{4}$
9. $6\frac{1}{2}$
10. $11\frac{2}{15}$
11. 9
12. $1\frac{5}{8}$
13. $1\frac{1}{2}$
14. $\frac{2}{3}$
15. 2
16. $\frac{1}{4}$
17. $\frac{2}{5}$
18. 3
19. $n = 11$
20. $n = 4$
21. $n = 12$
22. $n = 9$
23. $\frac{1}{5}$
24. $1\frac{1}{20}$
25. 55%

## GED Test Practice, *pages 142–144*

### Page 142

1. **(2) $\frac{1}{4}$** Choices (1), (3), and (4) are incorrect because they contain errors in reduction.
2. **(4) $\frac{3}{10}, \frac{1}{3}, \frac{2}{5}, \frac{5}{6}$** Choice (1) is incorrect because the fractions are listed from greatest to least. Choice (3) is incorrect because the fractions are listed from greatest to least numerators. Choice (2) contains an error in the order.
3. **(2) $7\frac{5}{6}$** Choice (1) is incorrect because it contains an error in addition. Choice (3) is incorrect because only the whole numbers were added. Choice (4) is incorrect because subtraction was used.
4. **(3) $\frac{1}{2}$ cup** Choice (1) is incorrect because multiplication was used. Choices (2) and (4) are incorrect because addition, the incorrect operation, was used.

### Page 143

5. **(4) $4\frac{7}{8}$ inches** Choice (1) is incorrect because the second fraction was inverted before multiplying. Choice (2) contains a multiplication error. Choice (3) used an incorrect operation, addition.
6. **(4) 12** Choice (1) is incorrect because the fractions were multiplied with inverting. Choice (2) used the incorrect operation. Choice (3) contains an error in reduction.
7. **(2) $2\frac{1}{5}$** Choices (1), (3), and (4) are incorrect because they contain errors in reduction.
8. **(3) $\frac{7}{10}$** Choice (1) is incorrect because the ratio is incorrect. Choices (2) and (4) contain errors in reduction.

### Page 144

9. **(3) 22.5 hours** Choice (1) is incorrect because the incorrect operation was used. Choice (2) is incorrect because it contains a multiplication error. Choice (4) is incorrect because an incorrect ratio was used.
10. **(3) 0.72** Choices (1), (2), and (4) are incorrect because the decimal points are in incorrect places.

## Mathematics Posttest, *pages 145–148*

### Page 145

1. **(1) 10,000** To round to the ten thousands place, look at the digit in the thousands place (0). Choice (2) is rounded to the hundreds place. Choice (3) is rounded to the tens place. Choice (4) is rounded to the thousands place.
2. **(4) 5,760** Multiply 480 by 12 to find the answer. In choice (1), the numbers were divided. Choice (2) used subtraction. Choice (3) added the numbers.
3. **(2) 32** Add 19 and 13 to find the total. In choice (1), the numbers were incorrectly added. In choice (3), the numbers were added without regrouping. The numbers were subtracted to get the answer in choice (4).

4. **(3) $35** Divide $175 by 5 to find the answer. In choice (1), the numbers were multiplied instead of divided. The numbers were added to get the answer for choice (2). In choice (4), the numbers were divided incorrectly.
5. **(4) $545** Subtract $1,375 from $1,920 to find the answer. Choice (1) is the sum of the numbers, not the difference. In choices (2) and (3), there are regrouping errors.
6. **(4) 6 thousands** In the number 6,947, the 6 is in the thousands place so the value is 6 thousands. Choices (1), (2), and (3) have incorrect place-value names.

**Page 146**

7. **(3) 1,280** Choice (1) is rounded to the nearest hundred, not ten. In choice (2), the number has been incorrectly rounded up to the next ten. In choice (4), the number has been rounded to the nearest thousand.
8. **(1) 1,319 < 1,561** Compare the digits in each number. Choices (2) and (3) have the < or > symbols incorrectly placed. Choice (4) is incorrect because the numbers are not equal.
9. **(4) 70 pounds** Multiply the numbers to find the answer. In choice (1), the numbers were subtracted. In choice (2), the numbers were added. Choice (3) shows an incorrect product due to a regrouping error.
10. **(3) $15** Divide the total fee ($60) by the number of sessions (4). Choice (1) is incorrect because the question asks for the fee for 1 session, and $60 is the total. In choice (2), 4 was subtracted from the fee ($60) instead of divided. In choice (4), the numbers were incorrectly divided.
11. **(3) $59** Subtract the amount of the rebate ($40) from the price of the phone ($99). In choice (1), the rebate was added to the price of the phone, not subtracted. Choice (2) is the price of the phone without a rebate. Choice (4) shows the rebate amount.
12. **(1) 748** Add 556 + 192 to find the answer. Choice (2) lists the number of workers the company started with. In choice (3), the numbers were subtracted. Choice (4) is the number of newly hired people.

**Page 147**

13. **(4) 21.4** Add the distances to find the answer. In choice (1), the distances were subtracted instead of added. In choice (2), only the whole numbers were added. In choice (3), mistakes were made in regrouping when adding.
14. **(1) $316.43** Subtract the deductions from the gross pay. In choice (2), there is a subtraction error. In choice (3), the numbers were subtracted without regrouping. Choice (4) is the sum of the numbers, not the difference.

**15. (3) $7.45** Divide the cost of the meal ($14.90) by 2 to find the answer. For choice (1), the price ($14.90) was multiplied by 2 instead of divided. In choice (2), 2 was subtracted from the price of $14.90. In choice (4), the division was not completed.

**16. (4) $1,272.00** Multiply $26.50 by 48 to find the answer. Choice (1) is incorrect because it is the sum of 48 and $26.50, not the product. In choice (2), the decimal point was improperly placed in the product. Choice (3) has a regrouping error.

**17. (3) 6 pounds** To estimate the total weight, round the weight of each item on the receipt ($2 + 1 + 1 + 2 = 6$). In choices (1) and (2), some numbers were rounded too low. In choice (4), the numbers were rounded too high.

**18. (4) 32** Divide $2\frac{2}{3}$ by $\frac{1}{12}$ to find the correct answer. Choice (1) is incorrect because the fractions were multiplied. In choice (2), the fractions were subtracted. In choice (3), the fractions were added.

**Page 148**

**19. (2) $55.50** To find 3% of 1,850, multiply 1,850 by 0.03. Choices (1) and (3) are incorrect because the percents were incorrectly changed to the wrong decimal (0.003 and 0.3). In choice (4), 1,850 was divided by 3.

**20. (4) $3\frac{3}{4}$ inches** Add $1\frac{1}{4}$ and $2\frac{1}{2}$ to find the answer. Choice (1) is incorrect because the mixed numbers were subtracted. In choice (2), the numbers were multiplied. In choice (3), the fractions were added incorrectly.

**21. (1) $2\frac{2}{5}$ inches** Subtract $1\frac{3}{10}$ from $3\frac{7}{10}$ to find how much more rain fell on Monday. Then reduce the fraction. Choice (2) is incorrect because just the whole numbers were subtracted. In choice (3), the numbers were added incorrectly. Choice (4) is incorrect because the numbers were added instead of subtracted.

**22. (1) 720** Solve the proportion $\frac{12}{1} = \frac{x}{60}$. Choice (2) is incorrect because it shows the amount packed in 30 minutes, which is a half hour. In choice (3), the numbers 60 and 12 were added. Choice (4) is incorrect because the terms in the proportion were in the wrong order.

**23. (4) $\frac{1}{18}$** Multiply $\frac{1}{3}$ by $\frac{1}{6}$ to find the answer. Choice (1) is incorrect because the fractions were divided. In choice (2), the fractions were added. Choice (3) is incorrect because the fractions were incorrectly multiplied.

**24. (2) $\frac{9}{24}$** Reduce all fractions to lowest terms to find the one equivalent to $\frac{3}{8}$. In choice (1), $\frac{8}{24} = \frac{1}{3}$. Choices (3) and (4) are both incorrect because the fractions both reduce to $\frac{1}{2}$.